PensareDiverso Publishing
“Quantum Landscapes” Series

Quantum Landscapes: Scientific and Inner Journeys.

Bruno Del Medico

Erwin Schrödinger
and the Paradox of Life.

Physics, diffuse consciousness, and Eastern philosophy.

Bruno Del Medico, Publisher.
PensareDiverso Editions
edizioni@delmedico.it
www.giardinoquantico.it

Table of Contents.

Introduction.

There is a cat locked in a box. Next to it is a device capable of releasing a poison if a radioactive atom decays. As long as the box remains sealed, according to an extreme interpretation of quantum laws, the atom can be in a superposition of states: decayed and not decayed. The cat would then, in a paradoxical and intolerable sense, be both alive and dead at the same time.

Few thought experiments have had the power of this image. It has found its way into textbooks, movies, and everyday conversations; it has become a metaphor for uncertainty, ambiguity, and even lives suspended in the face of a choice. But Schrödinger's cat was not invented to prove that the world is absurd, nor to suggest that human consciousness magically creates reality by observing it. It was conceived, in 1935, as a provocation: a way to show just how difficult it was to unquestioningly transfer the strange rules of the atomic world to the world of living beings.

Behind that box lies more than just a physics puzzle. There is an ancient and fundamental question: What is life? Erwin Schrödinger, one of the fathers of quantum mechanics, was not a physicist willing to leave this question on the margins of his discipline. His curiosity crossed boundaries that we today tend to regard as separate: matter and mind, biology and physics, scientific knowledge and philosophical reflection. In his famous book **What Is Life?**, published in 1944, he sought to understand how organisms manage to maintain their own order— —in a universe also governed by disorder. But Schrödinger was also a keen reader of Indian philosophy. In the Upanishads and in "Advaita Vedānta," he found expressions of an intuition that stayed with him throughout his life: the multiplicity of individual experiences might not be the final word on reality. Perhaps what we call "I" is a local perspective, a window opening onto a vaster whole. Perhaps consciousness is not an object among objects.

These are fascinating but delicate ideas. Science and Eastern philosophy speak different languages, ask different questions, and use different criteria to distinguish what is plausible from what is true. Yet they can meet on common ground: the recognition that reality does not always conform to the simplest images we construct of it.

This book portrays Schrödinger as the protagonist of an era in which matter ceased to appear solid, predictable, and completely separate from the observer. But the intellectual horizon opened up by Schrödinger extends even further, leading to a topic of current relevance: consciousness, one of humanity's greatest mysteries. Thus, the journey Schrödinger began in quantum physics extends into metaphysics, Eastern philosophy, biology, and neuroscience, all the way to the hypothesis of a unified field of consciousness, revealing an ever-deepening relevance in contemporary debate.

I. An imaginary animal, a real question.

"What we observe is not nature in itself, but nature as shaped by the way we ask questions."
Werner Heisenberg, *Physics and Philosophy*, 1958.

"Science, it seems, is completely incapable of telling us anything about red and blue, bitter and sweet, physical pain and physical pleasure; it knows nothing of beauty and ugliness, good and evil, God and eternity."
Erwin Schrödinger, *Nature and the Greeks and Science and Humanism*, 1954.

Schrödinger's cat is probably the most famous animal in the history of science, and certainly the most misunderstood. It has become a pop culture icon: it appears in comic books, TV shows, commercials, and social media jokes. Yet when Erwin Schrödinger conceived of it in 1935, he did not intend to make quantum physics more fascinating or mysterious. He wanted to challenge it.

Behind that closed box, behind the suspended fate of a cat tied to the decay of a radioactive atom, lies a question that extends beyond the confines of the laboratory: What can we say about reality before it is observed? And, above all, what happens when the laws devised to describe atoms and particles seem to extend to the world of bodies, animals, and life?

This chapter begins here. Not to transform quantum physics into a disguised form of spirituality, nor to deny the difference between a scientific theory and a philosophical reflection. On the contrary: we will try to carefully distinguish between these levels. Physics presents us with a rigorous problem, one whose implications remain open in some respects. Philosophy helps us understand why that problem concerns us. Finally, the question of life compels us to recognize that explaining the mechanisms of an organism does not yet amount to understanding what it means to be alive.

In 1935, in the pages of the German journal **Die* Naturwissenschaften*, Erwin Schrödinger proposed a thought experiment that would outlive its author. His article bore a seemingly technical title: *"The Current State of Quantum Mechanics*." It was not a work of fiction, nor was it an intellectual sleight of hand, nor even an attempt to argue that cats can truly be both alive and dead at the same time. Rather, it was a critique.

Schrödinger was forty-eight years old at the time. He was already one of the leading figures in the new physics:

in 1926, he had formulated the wave equation that bears his name, providing one of the defining mathematical expressions of quantum mechanics; in 1933, he had received the Nobel Prize in Physics, sharing it with Paul Dirac. But the very man who had contributed so profoundly to the development of quantum theory was unwilling to accept its most widespread interpretations without reservation. The point was not that the equations were wrong. They worked—and continue to work—with a precision that borders on the incredible. The point was something else: what, exactly, do those equations tell us about the world?

To bring this unease to light, Schrödinger devised a thought experiment. Imagine a closed box. Inside are a cat, a radioactive atom, a detector—a Geiger counter, as we would say today—a mechanism capable of triggering a hammer, and a small vial of hydrogen cyanide. If, within a certain time interval, the atom decays, the detector registers the event, the mechanism is triggered, the hammer breaks the vial, and the cat dies. If the atom does not decay, nothing happens, and the cat remains alive.

Quantum theory does not allow us to say with certainty, before the measurement, whether that particular atom has decayed or not. It can indicate a probability. In the case chosen by Schrödinger, after one hour, the probability is fifty percent: a fifty-fifty chance that the decay has occurred, and a fifty-fifty chance that it has not. So far, there is nothing particularly strange about this. Radioactivity is a real and well-studied phenomenon; quantum probabilities have been confirmed by countless experiments. The unsettling leap occurs when we link the fate of the atom to that of the animal.

If we take a certain description of quantum mechanics literally, the closed system seems to have to be represented as a superposition of possibilities: the atom having decayed and not having decayed; the mechanism having been triggered and not having been triggered; the cat being dead and the cat being alive. This is not simply a matter of our own ignorance, as when we do not know which card has been drawn from a shuffled deck. Here, the theory seems to suggest something more radical: before observation, the alternatives are part of the physical description itself.

It is at this point that Schrödinger writes: *"One can even construct entirely absurd cases.*" The cat is absurd because it brings an abstract difficulty out of the invisible realm of atoms and places it in the world we know: an animal, a box, poison, life, and death. This is precisely why the scene continues to strike us. Not because it's macabre, but because it forces us to confront a question we usually leave in the background.

When, exactly, does a possibility become a fact?

In everyday language, it's often said that Schrödinger's cat is "both alive and dead until someone opens the box." It's a memorable phrase, but it risks being misleading. It does not mean that a real cat can exist, by some mysterious power of consciousness, in an ambiguous biological state. Nor does it mean that the human eye has the magical ability to create reality simply by looking at it. The issue is more precise—and therefore more interesting.

In quantum physics, a system can be described by a combination of possibilities. An electron can be in a state that is not simply equivalent to a well-defined position of which we are unaware, but one that presents several measurable possibilities. When we perform a measurement, we obtain a specific result. The problem is understanding the transition from the description of possibilities to the single fact that is recorded: a trace on a screen, a number on an instrument, an event in a detector.

This is the so-called measurement problem. It is not an invitation to believe that "everything is possible"—a seductive notion but one foreign to physics. Quantum possibilities are defined with extreme precision by equations and experiments; they are not human desires, nor a repertoire of available miracles.

The cat serves to ask whether it really makes sense to extend the language of the subatomic world unmodified to everyday objects and living beings. Should we think that the entire box—atom, detector, poison, animal—remains in a sort of limbo until it is opened? And if so, what counts as an observation? The arrival of a human being? The Geiger counter? The molecules in the air? Light? The surrounding environment?

Werner Heisenberg, one of the fathers of quantum theory, wrote: *"What we observe is not nature in itself, but nature as it is revealed by the way we ask questions.*" This statement must be handled with care. It does not give us the right to say that a world independent of us does not exist. A detector does not function based on the opinions of the observer; a radioactive atom does not decay to please a consciousness. Heisenberg meant to say that physics never observes "reality in general": it observes phenomena through concrete apparatus, asking specific questions and obtaining specific answers.

Quantum physics has forced us to abandon a very reassuring image: that of a world made up solely of objects with predefined properties, like colored balls on a table, independent of any interaction. But it does not force us to conclude that reality is a private dream or that the human mind governs the cosmos. Between these two extremes—an overly " " materialism and a haphazard idealism—lies the challenging territory that the cat invites us to explore.

In the years following the paradox, physicists and philosophers have developed various responses. The Copenhagen interpretation, associated primarily with Niels Bohr, accepts a practical distinction between the quantum system and the measuring apparatus. The many-worlds theory, proposed by Hugh Everett in 1957, posits that the various possibilities do not disappear but come to pass in different branches of reality. Other hypotheses suggest that the collapse of the wave function—the transition from superposition to a determined outcome—is a real physical phenomenon and not merely an update to our description. Decoherence, studied in the 20th century by physicists such as H. Dieter Zeh and Wojciech Zurek, has also shown how quickly interaction with the environment renders large-scale superpositions fragile and practically unobservable.

These are different ideas, and not all of them enjoy the same level of consensus. None of them, however, justifies taking shortcuts. The fact that the issue remains debated does not mean that "science knows nothing"; it means that an extraordinarily effective theory continues to pose serious problems regarding how to interpret its own success.

Schrödinger's cat is not a treatise on biology. However, its power also stems from the fact that, inside the box, there is not just an object: there is a living being.

A rock can break, a clock can stop, a star can collapse. But a cat is not merely a certain amount of organized matter. It breathes, metabolizes, perceives, reacts, and suffers; it is part of an evolutionary history spanning billions of years. Its life is not a sentimental detail added to the scene: it is what makes the neutrality of the paradox intolerable.

This reaction does not prove that life must defy the laws of physics. Everything that happens in an organism also occurs through physical and chemical processes: energy exchanges, molecular reactions, electrical signals, transformations of matter. It would be futile to seek a suspension of the laws of nature in living beings.

But acknowledging this does not settle the question. To say that life is entirely physical is not the same as saying that we have already clarified what life is.

In 1943, during lectures he gave in Dublin at the *"Dublin Institute for Advanced Studies,"* Schrödinger addressed precisely this question. From those lectures emerged, in 1944, **What Is Life?**, one of the most influential books of the twentieth century in the dialogue between physics and biology. The book did not provide a definitive definition of life. However, it brought one problem into sharp focus: How do organisms maintain an ordered form over time, reproduce, and transmit hereditary instructions, even though they are composed of matter subject to disorder and wear and tear?

Schrödinger used the term "aperiodic crystal" to envision a material structure capable of storing a large amount of information without the regular repetition found in an ordinary crystal. He did not discover DNA: its double-helix structure would not be described by James Watson and Francis Crick until 1953, thanks in part to the decisive X-ray diffraction data obtained by Rosalind Franklin and Raymond Gosling. But Schrödinger's book helped make it conceivable that heredity had a specific physical and molecular basis. Francis Crick later acknowledged how much that reading had influenced his decision to switch from physics to biology.

Here is an important first lesson. The big questions do not belong to a single discipline. Sometimes a question posed by a physicist prompts a biologist to look for a bet ; sometimes a biological discovery forces philosophy to reexamine its own images of humanity.

Darwin concluded **On the Origin of* Species* with words that have remained famous: *"There is grandeur in this view of life."* It was not a consolatory phrase. It was the recognition that life, viewed through the lens of evolution, does not lose its splendor: it gains it in a new dimension, one made up of deep time, transformations, fragility, and continuity among all species.

The cat in the box belongs to this very same continuity. It is not a symbol of humanity; it is not a spiritual metaphor disguised as an animal. It is a living being, and precisely for this reason it reminds us that our theories, when they touch the world, also touch something that can be hurt, that is born and dies.

A thinker who defied convention.

Schrödinger was neither a secular saint of science nor a spiritual master to be placed outside of history. He was a complex man, at times contradictory, with a private life that need not be idealized. But he was also one of those rare scientists who do not shy away from the big questions for fear of appearing unprofessional.

Born in Vienna in 1887 and raised in Central European culture, he lived through some of the defining upheavals of his century. He worked in Zurich and Berlin; in 1933, with the rise of Nazism, he left Germany. After a stint at Oxford and several more years marked by European instability, he found a relatively secure workplace in Dublin. Irish Prime Minister Éamon de Valera, a mathematician by training and a central figure in Irish politics, had promoted the founding of *the "Dublin Institute for Advanced* Studies" precisely to provide a space for high-level research. It was there that Schrödinger wrote his most famous book, **What Is Life?**. His curiosity, however, extended beyond the natural sciences. He read Greek philosophy, pondered the relationship between mind and matter, and studied with interest the Upanishads and "Advaita Vedānta," the Indian tradition that emphasizes the profound unity of reality and challenges the absolute separation between the individual self and the whole.

Here we must be very clear. The Upanishads did not anticipate quantum mechanics. They do not contain atomic theory, Schrödinger's equations, or particle physics. Seeking direct correspondences between ancient spiritual intuitions and modern experimental results almost always yields vague analogies and unprovable claims.

Yet, it would be equally superficial to ignore that Schrödinger found in those traditions a language suited to expressing his own personal unease: the difficulty of conceiving of consciousness as a mere object among objects and the difficulty of accepting that the world is truly composed of totally separate realities.

"*The world is given to me only once, not as an existing world and another perceived world. Subject and object are one and the same,*" he would write in *Mind and Matter*. This is not a conclusion of quantum mechanics. It is philosophy. It is a meditation on the fundamental fact that every description of the world is always formulated by someone who, in turn, is part of that world.

The distinction is crucial. Here, physics speaks: quantum systems, probabilities, measurements, repeatable experiments. Here, interpretation comes into play: what picture of reality can we derive from those results? Here, finally, we enter a more existential and metaphysical plane: what does it mean to be a consciousness in a universe that we can investigate but not fully possess?

Confusing these levels impoverishes both science and philosophy. Keeping them distinct does not mean isolating them in watertight compartments; it means allowing them to engage in dialogue without forcing them to pretend to be the same thing.

Schrödinger's cat remains in its box not because it awaits our definitive answer, but because it teaches us not to settle too quickly for easy answers. It does not tell us that consciousness creates matter. It does not tell us that physics confirms ancient wisdom. Nor does it tell us that reality is incomprehensible. It tells us, more soberly and more profoundly, that the world is perhaps less like the images we have come to rely on for reassurance. And that, if we wish to approach the question of life, we must patiently navigate through physics, biology, philosophy, and the silence that remains when each of them has said all it can say.

Why "simultaneously" alive and dead? Quantum superposition.

When we say that, in quantum mechanics, something is "simultaneously alive and dead," we are not describing a normal situation in the everyday sense of the words. It is a provocative way—coined with Schrödinger's famous cat paradox—to express a real and experimentally verified characteristic of quantum systems: superposition.

In the ordinary world, we're used to thinking that things always have a precise state, even when we don't know it. If a coin is covered by a hand, for example, it has already landed heads or tails: the uncertainty is ours alone. We do not know which side is facing up, but the coin nonetheless has a definite outcome. In the quantum world, at least according to the standard description of physics, the situation is more radical. Before a measurement, a particle may not have a definite property in the classical sense; its state may contain multiple possibilities simultaneously.

However, we must not imagine a particle as a normal little ball that is physically split into two halves. Quantum mechanics describes the particle through a wave function—that is, a mathematical description that encompasses the possibilities of the obtainable outcomes. As long as the system is isolated and not subjected to a measurement that distinguishes between these possibilities, they remain present in the description and can even influence one another. This effect, called interference, is the most important indication that this is not simply a matter of our own ignorance.

The double-slit experiment makes all of this more intuitive. Imagine a source that emits electrons—or photons—one at a time. In front of the source is a barrier with two very narrow slits, one on the left and one on the right. Behind the barrier is a sensitive screen, on which each electron or photon leaves a small mark at the point where it strikes.

If only one slit is open, the behavior seems fairly straightforward: the particles pass through the opening and strike the screen, forming a certain pattern. If the other slit is opened, a similar pattern is obtained, shifted to the corresponding side. It would be natural to think that, by opening both slits, the final result would simply be the sum of the two groups of particles: some passing to the left and others to the right.

But that is not what happens if we do not try to determine the path. Instead, an interference pattern appears on the screen: areas where a great many particles arrive alternate with areas where very few or almost none arrive. It is the same type of pattern one would expect from waves coming from the two slits and overlapping: at certain points they reinforce each other, at others they cancel each other out.

The most surprising thing is that this pattern emerges even when the particles are sent one at a time. Each electron arrives on the screen as a single localized spot, not as a diffuse smear; but as a vast number of these spots are collected, they gradually form the interference fringes. It is as if the single particle had passed through both slits, and its behavior had been influenced by the presence of the same particle emerging from both slits.

To say that "the particle passes through both slits simultaneously" is a useful simplification, provided it is understood correctly. More precisely, before a measurement of the path is made, the particle's quantum state is a superposition of the possibility of "passing through the left slit" and the possibility of "passing through the right slit." It is not correct to necessarily imagine two tiny, perfectly separate material copies, as in a science fiction story. It is more accurate to say that the theory does not assign the particle to just one of the two paths, and that the two possibilities remain coherent—that is, capable of interfering with one another.

At this point, the experiment can be modified: a detector is placed near the slits, designed to determine which slit each particle passes through. When this is done, the interference pattern disappears. What remains on the screen is a distribution consistent with the classical view: each particle is found to have passed through either the left slit or the right slit. We never find the detector registering both paths in a single event. Every measurement yields a definite result.

In the traditional version of quantum mechanics, it is then said that the wave function has "collapsed." Before the measurement, the system was described by the superposition of the two paths; after the measurement, it is described by a single, actually observed outcome. The collapse does not mean that the human eye has magically created reality. The word "observation," in this context, can be misleading: in physics, it does not primarily mean looking with one's consciousness, but rather causing the system to interact with something capable of distinguishing and recording the alternatives.

The reason the measurement destroys the interference is that the particle is no longer isolated. If it passes to the left, the detector assumes a state containing the "left" information; if it passes to the right, it assumes a different state, containing the "right" information. The particle and the detector thus become closely correlated. Once this information exists in the detector—and it often disperses immediately into the environment, into the air, light, heat, and billions of surrounding particles—the two alternatives can no longer interfere with each other as they did before.

This is where Schrödinger's cat comes into play. Schrödinger did not invent the paradox to claim that cats are truly in a mysterious state between life and death; on the contrary, he proposed it to show how problematic it is to take quantum superposition literally when extending it to large objects in everyday life.

The thought experiment imagines a closed box containing a cat, a radioactive substance, a detector, and a mechanism connected to a vial of poison. If a radioactive atom decays, the detector registers it and triggers the mechanism that breaks the vial: the cat dies. If the atom does not decay, nothing happens and the cat remains alive. After a set interval of time, the atom may have, for example, an equal probability of having decayed or not having decayed.

A radioactive atom is a quantum system. If, before measurement, we describe it as a superposition of "decayed" and "not decayed," then everything causally linked to the atom is included in that same description: on the one hand, there is the possibility "atom decayed, detector triggered, poison released, cat dead"; on the other, there is the possibility "atom not decayed, detector not triggered, poison not released, cat alive."

This is the sense in which we speak of the cat as "both alive and dead at the same time." It does not mean that, inside the box, there is a cat that can be perceived as half-alive and half-dead, nor that biology ceases to function. It means that, if we apply the rules of quantum mechanics without modification to the complete and perfectly isolated system, the description simultaneously contains the two correlated possibilities.

Opening the box is analogous to taking a measurement. Before the box is opened—according to the simplest version of the story—the system is represented by the superposition of the two alternatives. When the box is opened and the result is checked, only one fact is found: the cat is either alive or dead. That is, "collapse" has occurred, and the cat transitions from a probabilistic state to a certain state. In the traditional description, the act of measurement selects one of the possible outcomes, and the superposition "collapses" into that observed reality.

However, we must be very cautious here. In the case of the double-slit experiment, interference is actually observed in the laboratory and directly confirms the existence of quantum superposition for particles, atoms, and—using highly sophisticated techniques—increasingly larger systems. In the case of a real cat, however, the situation is very different. A cat interacts incessantly with its environment: it emits heat, breathes, is exposed to radiation, exchanges molecules with the air, produces sounds and movements, and interacts with the box itself. Each of these interactions leaves traces of its state.

This process is called decoherence. If there were two components—"live cat" and "dead cat"—they would become entangled almost immediately with a vast number of details in the environment. The environment would end up containing, in effect, the information about which of the two situations has occurred. At that point, the interference between "alive" and "dead" would become inaccessible, and the " " would be practically impossible to observe. For a macroscopic object, this happens in an extremely short time.

Decoherence explains why we do not see tables in two places at once, people overlapping between two rooms, or cats producing interference between life and death in our daily lives. Quantum behavior does not disappear in the macroscopic world: all objects are made of quantum particles. But their interactions with the environment are so numerous and rapid that observable quantum coherence vanishes.

A profound question remains: Does collapse truly occur as a fundamental physical phenomenon, or is it merely an effective way of describing the fact that, after a measurement, we observe a definite outcome? Physics does not yet have a single answer accepted by everyone.

In the most traditional interpretation, collapse is an essential part of the description: the measurement transforms a superposition into a concrete result. In other interpretations, such as the many-worlds interpretation, no collapse occurs: both alternatives continue, but become separate branches that can no longer interfere with one another. In one branch, an observer opens the box and finds the cat alive; in another branch, a different observer, corresponding to the first, opens the box and finds the cat dead. This interpretation is well-known, but it is no more proven than the other interpretations: it is simply a different way of understanding the same formalism and the same experimental data.

The fundamental lesson of the double-slit experiment is therefore this: a quantum particle should not always be thought of as a classical object with pre-determined and hidden properties. Before the path is measured, the quantum description contains both possibilities, and the phenomenon of interference demonstrates that these possibilities have real effects. When information about the path is obtained, the system no longer exhibits that interference, and each individual measurement yields a precise outcome.

Schrödinger's cat takes this idea to the extreme to highlight its paradoxical nature. It does not tell us that real cats normally lead a double existence; rather, it forces us to ask where the quantum world of superimposed possibilities ends and where the everyday world of definite facts begins. The experimental answer is that the transition depends primarily on interaction with the environment and decoherence. The philosophical answer as to what “really” happens, however, remains one of the most fascinating questions in modern physics.

The cat as an emblem of the problem of measurement and observation.

Few scientific images have had a cultural impact comparable to that of Schrödinger's cat. It has appeared in novels, comic books, television series, and newspaper headlines; it has been turned into a recurring joke, a metaphor for undecided relationships, uncertain elections, and even medical diagnoses awaiting confirmation. To say that someone is "like Schrödinger's cat" has become a way—often imprecise but eloquent—of saying that a situation remains suspended between two possibilities.

And yet, at the outset, there was no desire whatsoever to come up with a brilliant idea. On the contrary, it was a protest. In 1935, Erwin Schrödinger published an article titled *"Die gegenwärtige Situation in der Quantenmechanik"* in three consecutive installments in the German journal **Die* Naturwissenschaften*. In those pages, written while he was teaching at the University of Oxford, he sought to bring to light a difficulty that seemed intolerable to him: if we take the rules of quantum physics literally, how far can we extend them? Do they apply only to atoms, electrons, and photons? Or must they also apply to everyday objects, laboratory instruments, animals, and ourselves? It is in the third article of the series that Schrödinger's famous thought experiment involving the cat appears.

Schrödinger introduces his example with an almost theatrical tone. *"One can even construct entirely burlesque cases,"* he writes. But that "burlesque" quality must not lead us into deception. The paradox is comical only on the surface; at its core, it is one of the most serious questions that twentieth-century physics has bequeathed to us.

The decisive point is that, in the quantum description, prior to measurement, the atom is not treated simply as having already decayed or not yet decayed, even if we do not know which of the two has occurred. Its state is expressed as a superposition of possibilities. This term, "superposition," must be handled with care. It does not mean that the atom is in a state of confusion because we lack information, like a face-down card on a table. It means that the theory describes it through a combination of possible outcomes—a combination that can produce very concrete observable effects, such as the interference patterns obtained with electrons or photons.

As long as we remain in the realm of particles, we can accept this strangeness almost with resignation. The microscopic world doesn't resemble our own, we tell ourselves; it's counterintuitive, but experiments confirm the equations with extraordinary precision. The problem arises when the fate of the atom is linked to something large, visible, and irreducibly familiar: a living animal.

If the state of the atom is a superposition of decay and non-decay, and if the cat's fate depends on decay, then the entire system—atom, counter, mechanism, vial, and cat—also seems to have to be described as a superposition: a live cat and a dead cat. Not because the poor animal is in an ambiguous medical condition, nor even because we do not yet know what has happened. The point is more radical: the theory, applied without exception, seems not to allow us to choose a single outcome before a measurement takes place.

This is where the crack appears. In ordinary life, things either happen or they don't. A glass either falls or stays on the table. A letter is either mailed or remains in the drawer. A door is either open or closed. Of course, we may be uncertain: we may not have seen the glass, checked the door, or received the letter. But our ignorance does not seem to alter what has happened.

Schrödinger's cat challenges precisely this certainty. It doesn't simply ask us, *"What do we find when we open the box?"* It asks us: when, exactly, does a possibility become a fact?

Has the Geiger counter already "observed" the atom? Is the mechanism that triggers the count already a measurement? Does the broken vial constitute a definite event? The cat itself, if it is alive, is it not a system capable of reacting to what happens to it? And if all this is not enough, must we wait for a human being to open the box? But then, what do we really mean by words like "observation" and "measurement"?

It's important to pause for a moment, because a great deal of confusion has arisen on this point. Quantum physics does not simply and unanimously assert that human consciousness "creates" reality by observing it. This claim, widespread in many spiritual discourses and sometimes in certain hasty popularizations, is too strong. In the laboratory, a measurement can be performed by an automatic detector; no one needs to be staring at an instrument's dial at the exact moment the data is recorded. A photographic plate, an electronic sensor, or a bubble chamber can record traces of microscopic events even in an empty room.

But this is precisely where the problem becomes more subtle. If a device records an event, that device is made of atoms. And atoms, at least in principle, obey the same quantum laws as the initial atom. Where, then, does the boundary lie between the world of possibilities and that of facts? Between the quantum description and everyday experience, in which a needle points to a precise value, a photograph shows a point in a specific location, and a cat is either alive or dead?

Niels Bohr, one of Schrödinger's great interlocutors and adversaries, tended to respond pragmatically. To discuss an experiment, he argued, we must describe the apparatus and the results using ordinary language—the language that allows us to say where a detector is located, what number appears on a screen, and what trace has been observed. Not because the everyday world is naively separate from the quantum world, but because without results that can be expressed and shared, physics as a human practice does not even exist.

Schrödinger was not satisfied. His cat precisely represents his impatience with a separation that seems necessary in language but is difficult to justify within the equations. If the theory is universal, why should it stop at a measuring device? If it is not universal, where and why does it cease to apply?

Albert Einstein sensed a similar difficulty, though he expressed it using different imagery. He did not like the idea that physical reality remained incomplete or indeterminate until the moment of measurement. His dissent did not stem from nostalgia for a mechanical and predictable universe: rather, it was a demand that the world possess its own consistency, independent of whether anyone questioned it. Schrödinger, Einstein, and Bohr were not naive defenders of a common sense undermined by modernity; they were the protagonists of a dizzying debate, because they knew that the new physics worked, but they disagreed on what it authorized us to say about reality.

The cat paradox, therefore, does not prove that "everything is relative," nor that the mind can arbitrarily decide the fate of things. It demonstrates something more precise and more unsettling: quantum equations successfully describe the evolution of possibilities, but it remains difficult to understand how, in concrete experience, a single outcome emerges.

Throughout the twentieth century, various answers were proposed. Some interpretations maintain the distinction between the quantum system and the measuring apparatus; others hypothesize that the transition to a definite result is a real physical process, linked to some form of wave function collapse; still others suggest that all outcomes compatible with the equations occur in different branches of reality. This last idea, formulated in 1957 by the American physicist Hugh Everett III, is known to the general public as the "many-worlds" hypothesis. From this perspective, there would not be a single cat that mysteriously shifts from a state of limbo to life or death: there would be different outcomes, each linked to a different history of the observer.

It is a powerful, almost novelistic image, and for this very reason it should be viewed with caution. It is not an experimentally established fact that the universe branches into parallel worlds in the popular sense of the term. It is an attempt to interpret the equations themselves without introducing a special boundary between the microscopic and the macroscopic. Other interpretations take different paths. Physics, in this regard, offers solid results; their ultimate meaning remains open to discussion.

An important contribution has come, since the second half of the twentieth century, from the study of decoherence. The German physicist Heinz-Dieter Zeh formulated the initial ideas in 1970; Wojciech Zurek then developed them in a decisive way. The fundamental idea is that a real system never exists in perfect isolati . It continuously interacts with its environment: air molecules, thermal radiation, vibrations, light. These interactions make superposition extremely fragile when it involves large objects. A cat, a vial, or a Geiger counter cannot be treated as isolated particles in a perfect vacuum.

Decoherence explains why, in practice, we do not see cats in superposition and why the strangest quantum behavior becomes nearly impossible to observe on an everyday scale. But it does not answer every question. It explains very well why the alternatives—alive cat and dead cat—cease to interfere with one another; it does not resolve the question of how a single alternative becomes the reality we experience. It is a subtle but decisive difference. Saying that a problem has been clarified does not always mean that it has been eliminated.

Perhaps this is why the cat continues to haunt us. It is not merely a thought experiment in modern physics. It is an image that touches on a deep-seated habit of ours: the belief that the world is already completely defined before any encounter with it. Quantum mechanics does not necessarily force us to abandon this idea in every possible sense; it does, however, force us to examine it more rigorously.

When we imagine opening the box, we encounter more than just an imaginary cat. We encounter a question about the relationship between what can happen and what actually happens, between the language of equations and the world of experience, between the data recorded by an instrument and the meaning we attribute to it.

For now, there is no need to turn this question into a philosophy of consciousness, nor to use it to claim that the universe depends on our thoughts. That would be jumping the gun. It is enough to recognize the significance of the unease that Schrödinger had brought to light: if physics manages to describe matter down to its most intimate structures, why does it become so difficult to explain the simple occurrence of a fact?

The box remains closed not because a definitive answer is lacking, but because it forces us to carefully distinguish between what we know, what we interpret, and what, for the moment, we can only question.

What distinguishes a cat from a mechanism?

It's obvious, but also important to emphasize: the famous "Schrödinger's cat" was never an experiment actually carried out, nor did Schrödinger propose actually locking an animal in a box with poison and radioactive material. It was a thought experiment—that is, an imaginary scenario devised to highlight a conceptual difficulty in quantum mechanics.

The implicit, unsettling question is: what is life, really? Schrödinger's cat strikes us not only because it is involved in a cruel and absurd thought experiment, but because it is not a ball, a needle, or a fragment of metal.

The cat is a living being. It breathes, feels, reacts, and possesses a unique biological history. And it is precisely this presence—seemingly incidental—that shifts the problem beyond the confines of laboratory physics.

If there had been only a detector in the box—either on or off—the paradox would have retained all its technical complexity. But a cat that is either alive or dead introduces something that none of us can regard merely as an experimental datum. It introduces vulnerability, time, the organism, the concrete difference between a configuration of matter and a creature that experiences hunger, fear, sensory memory, and a heartbeat. In short, it introduces life.

It is then that the question of quantum measurement broadens. It no longer concerns only the moment when an instrument produces a definite result. It asks us whether a living can be fully understood using the same language with which we describe a crystal, a cloud of gas, or an elementary particle. Not in the trivial sense that biology should reject physics: every cell, every protein, every nerve impulse obeys physical laws. The question is another. Is it enough to know the material elements and their interactions to understand what it means to be alive?

Here we must distinguish between these levels without confusing them. On the level of science, life is not a mysterious substance added to matter. There is no "vital force" separate from the laws of nature that modern biology has had to introduce to explain the growth of a plant, the healing of a wound, or the development of an embryo. Living beings are made of the same elements found throughout the rest of the universe: carbon, hydrogen, oxygen, nitrogen, phosphorus, and many others. The iron in hemoglobin that carries oxygen in our blood is the same chemical element found in rocks and stars. The calcium in our bones, the sodium that enables the transmission of nerve impulses, and the water that makes up most of our bodies hold no biological secrets when considered individually.

Yet, when these elements become part of a cell's structure, something remarkable happens. Not because matter ceases to be matter, but because it organizes itself in a way that allows it to sustain itself, transform, repair itself, reproduce, and pass on information across generations.

A rock can last for millennia, but it does not actively preserve itself. It does not need to seek energy to remain as it is. A flame consumes energy, grows, and spreads, but it does not inherit a set of instructions capable of changing and evolving. A crystal can generate ordered structures, but it does not build a dynamic boundary between itself and its environment; it does not metabolize, nor does it modify its behavior o s in light of a long history of adaptations. A living being, on the other hand, is an unstable equilibrium that must be continually renewed.

Consider a simple organism such as a bacterium. It has no face, no thoughts in the human sense of the term, and no nervous system. And yet, through its chemical receptors, it recognizes certain environmental conditions; it moves toward nutrients and away from harmful substances; it repairs damage to its own genetic material; it divides. That tiny cell is not separate from the world: it lives precisely because it exchanges matter and energy with the world. But it does not dissolve into it. It maintains a fragile distinction between an inside and an outside.

This distinction is one of the keys to life. The French biologist and philosopher Georges Canguilhem, a reclusive yet pivotal figure in twentieth-century scientific thought, wrote that life is "dynamic polarity": living beings do not simply endure their surroundings, but distinguish, so to speak, between what is favorable to them and what threatens them. We must not attribute conscious intentions to a cell. That would be a mistake. We can, however, recognize that an organism is not a passive object: it has its own way of staying alive, of compensating for imbalances, of coping with variations and damage.

A thermometer records temperature, but it does not care about the temperature. An organism regulates its own temperature—when it can—because the continuity of its vital processes depends on maintaining it. This simple yet immense difference brings us closer to the question: What is life?

Erwin Schrödinger did not invent this question; it is as old as philosophy and medicine. But he had the merit of reformulating it at an extraordinary moment in history. In February 1943, in the midst of World War II, he gave a series of public lectures at Trinity College in Dublin dedicated to the relationship between physics and biology.

Ireland had remained neutral in the conflict, and Dublin offered Schrödinger—an Austrian who had already been forced to leave Central Europe as it was overrun by Nazism—a relatively stable workplace at the *"Dublin Institute for Advanced Studies,"* founded in 1940.

From those lectures emerged, in 1944, a short book destined to have an enormous influence: *"What Is Life?"*

The title sounds almost disarming. It does not promise a new theory of the cosmos; it does not herald a revolution. And yet it tackles a question that has stumped biologists, philosophers, doctors, and poets alike: How does matter manage to maintain a form, reproduce a structure, and leave a legacy?

Schrödinger did not offer a definitive definition of life. It would be misleading to read his work as though he had solved the problem. His contribution was primarily to highlight two aspects that seemed essential to him. The first concerned order. An organism, even while living in a universe where energy tends to disperse and differences tend to fade, manages temporarily to maintain an extraordinarily complex structure. The second concerned heredity: how can physical matter preserve and transmit—with remarkable fidelity but also with the possibility of changes—the instructions necessary for the construction of an organism?

To make the first question more intuitive, Schrödinger used a formula that has since become famous: the organism, he wrote, *"feeds on negative entropy."*

The second law of thermodynamics states that, in an isolated system, entropy tends to increase: energy is dispersed, and spontaneous processes move toward more disordered states. The main consequences are:

- Heat flows spontaneously from hot to cold, not the other way around.

- Not all heat can be converted into useful work: every engine inevitably has losses.
- Natural processes have a direction: a glass can break, but it cannot reassemble itself. Energy tends to become less usable: it remains present, but is more dispersed.
- This explains the "arrow of time": we distinguish the past from the future because total entropy increases.

In practice, everything moves from order to disorder.

Today, biologists and physicists prefer to express themselves in more precise terms: organisms maintain their organization by absorbing energy and matter from the environment and expelling heat and waste products. They do not violate the second law of thermodynamics; on the contrary, they live precisely through ceaseless exchanges that increase the overall disorder around them.

There is nothing magical about this. A plant uses the energy of sunlight to build complex molecules; an animal derives energy from the transformation of other molecules; a cell relentlessly renews its own components. The body, which we consider a relatively stable entity, is, in reality, a river of exchanges. We remain ourselves not because the matter that composes us is immobile, but because a form of organization persists while the matter passes.

This is physics applied to life. But it is also the point at which physics, on its own, leaves us with unanswered questions. It can explain the energetic constraints within which an organism exists; it does not necessarily exhaust the human meaning of the experience of being alive. It can describe the mechanisms of breathing; yet it does not, for that reason, tell us what a person feels when they sense their breath slowing by the sea or quickening during a farewell. It can clarify many of the material conditions of memory; it does not automatically transform a memory into the equation that describes some of its processes.

To explain is not to diminish. Explaining how a rose works does not erase its scent; nor does the scent replace botany. Intellectual maturity perhaps consists precisely in not forcing one form of knowledge to do the work of another.

The second major theme of *"What Is Life?"* concerned heredity. Schrödinger imagined that the physical basis of hereditary traits must be a highly ordered molecular structure, though not periodic like a common crystal. He called it an "aperiodic crystal": a form stable enough to store information, complex enough to contain a wide variety of instructions.

The image was not a detailed prediction of the DNA double helix, which would be proposed nearly ten years later. But it had remarkable guiding power. Many young scientists realized that the secret of heredity could be sought not in some vague essence of life, but in the concrete structure of molecules.

Among those who read that little book with particular attention was Francis Crick. In his scientific autobiography, **What Mad* Pursuit* (1988), Crick recalled that reading Schrödinger's work played a decisive role in convincing him that biology could become the focus of his research. In 1953, in Cambridge, Crick and James Watson presented the double-helix model of DNA in *Nature*, drawing in part on X-ray diffraction data obtained by Rosalind Franklin and Raymond Gosling at King's College London and on the work of Maurice Wilkins.

It is incorrect to portray this story as if Schrödinger had "predicted DNA." The history of science is almost always more collaborative and less linear than its myths suggest. But it is fair to acknowledge that **What Is Life?** helped build an intellectual bridge: it showed many physicists that the problems of biology were not a minor or confusing field, but rather a frontier capable of demanding imagination, rigor, and courage.

One important fact remains, however. After DNA, after molecular biology, after the ability to sequence entire genomes, the question "What is life?" has not disappeared. If anything, it has become richer and more demanding.

We can list useful properties: metabolism, reproduction, evolution, response to the environment, and the ability to maintain an organized structure. But every list encounters borderline cases. Viruses, for example, possess genetic material, evolve, and are subject to natural selection; however, they lack an autonomous metabolism and depend on the cells they infect to replicate. Are they alive? Many biologists prefer to consider them biological entities on the edge of life, rather than living organisms in the full sense. This is not merely a matter of semantics: it is a sign that nature does not always conform to the clear-cut categories that our language would like to impose on it.

Even the origin of life on Earth remains, to a large extent, an open question. We know that our planet formed about 4.54 billion years ago and that the oldest evidence of biological activity dates back to at least 3.5 billion years ago—perhaps even further back, according to some interpretations that are still debated. But we do not know with certainty the sequence of steps through which prebiotic chemistry gave rise to the first systems capable of replication and evolution.

This uncertainty is not a failure of science. It is science's honest way of indicating where research continues. Knowing that we do not yet have a complete answer does not give us the right to fill the void with any consolatory explanation. Nor does it oblige us to pretend that the mystery is merely a temporary label for something that will soon be reduced to a diagram.

On an existential level, life concerns us before any definition. We are beings who seek to understand life while already immersed in it. We study cells with brain cells; we formulate theories using a language that emerged over the course of our evolutionary history; we observe nature as one of its later, more reflective forms. We are not external spectators facing a distant object. We are a part of life that, for a very brief stretch of its history, has become capable of asking itself where it comes from and what it is.

Charles Darwin concluded **On the Origin of* Species*, published in 1859, with words that retain their full power: *"There is something grand in this view of life."* It was not an invitation to naturalistic sentimentality. It was the recognition that life, viewed in the depth of time, appears as a tapestry of transformations, lineages, extinctions, and new forms. Darwin wrote: *"From a single beginning, endless beautiful and marvelous forms have been and are evolving."*

Schrödinger, starting from a very different question, leads us toward a similar boundary. His cat is not merely the emblem of a physical theory grappling with the act of measurement. It is also a reminder—perhaps unintentional but inevitable—not to treat living beings as a marginal detail. Inside the box there is not a complicated object: there is an organism. And an organism is matter that, for a certain time, manages to maintain a form, to exchange energy with , and to defend a fragile continuity against dissolution.

What this continuity ultimately is—and whether it can be fully described in the language of physics—is a question we should not close too soon. For now, we can take a more modest, yet perhaps more fruitful, step: recognizing that life is not an exception to the laws of nature, and yet it is the place where those laws intertwine in a way that continues to surprise us.

A thinker who defied convention.

Erwin Schrödinger was not merely the name associated with a fundamental equation of modern physics and a cat that, despite itself, became a universal celebrity. He was a figure harder to pin down: a rigorous physicist, a Nobel laureate, capable of making a decisive contribution to the birth of quantum mechanics, but also a man who never fully accepted the idea that science should limit itself to calculating results and remain silent on the human meaning of the world it describes.

In an era rapidly moving toward specialization, Schrödinger retained an almost old-fashioned disposition: he wanted to understand the whole. Not in the naive sense of someone who demands a definitive answer to every question, but in the more demanding sense of someone who refuses to separate matter from life, life from the mind, and the mind from the question of our place in the universe too soon.

Born in Vienna in 1887, Schrödinger grew up in the Austro-Hungarian capital as the city was transforming into a laboratory of modernity: Mahler was renewing its musical world, Klimt was shaking up its visual imagination, and Freud was paving the way for the exploration of the unconscious. It was at that crossroads of science, art, and philosophy that the cultural horizon of his generation took shape. For him, Vienna was not merely a place of birth; it was a spiritual atmosphere. In the Central Europe of his formative years, the disciplines were not yet entirely confined to their own spheres. A physicist could read Goethe, a doctor could discuss philosophy, and a mathematician could take an interest in language and literature.

Schrödinger belonged to that cultured, cosmopolitan, and fragile world that World War I and then Nazism would shatter. During the war, he served in the Austro-Hungarian army, primarily in artillery units stationed on the Italian front. It is a detail worth recalling: before becoming one of the great names in twentieth-century physics, Schrödinger had experienced firsthand industrial warfare, the turmoil of history, and the fall of an empire.

His most brilliant scientific period came in Zurich. In 1926, while serving as a professor of theoretical physics at the University of Zurich, Schrödinger published a famous series of six articles in *the *Annalen der Physik**, titled *"Quantisierung als Eigenwertproblem"* (*"Quantization as an Eigenvalue* Problem").

In those works, he developed the wave theory of quantum mechanics and introduced the equation that bears his name. The initial idea was inspired by Louis de Broglie's hypothesis: if light—traditionally thought of as a wave—could also manifest itself as a particle, perhaps material particles—such as electrons—possessed wave-like properties. Schrödinger therefore sought a mathematical law capable of describing these "matter waves." He found it in a differential equation that determines how a wave function—usually denoted by the Greek letter ψ (*psi*)—changes in space and time.

When applied to the hydrogen atom, the equation reproduced with remarkable precision the energy levels already observed experimentally and explained why the atom emitted or absorbed light only at certain frequencies. The quantized orbits introduced by Bohr were thus replaced by stationary states: permissible wave configurations, analogous—only intuitively— —to the possible vibrations of a musical string. Quantization no longer appeared as a rule imposed from the outside, but emerged from the mathematical conditions imposed on the wave function.

Schrödinger also demonstrated that his wave mechanics was mathematically equivalent to the matrix mechanics developed shortly before by Werner Heisenberg, Max Born, and Pascual Jordan. The two formulations seemed very different: one made use of waves and continuous quantities, the other of matrices and discrete quantities. Yet they led to the same observable predictions. This equivalence contributed decisively to the consolidation of the new quantum mechanics.

The physical meaning of the wave function, however, was not immediately apparent. Schrödinger initially hoped to interpret it as a real distribution of matter or charge; shortly thereafter, Max Born proposed the probabilistic interpretation: the square of the modulus of the wave function expresses the probability of finding a particle in a certain region of space. This marked one of the most profound breaks with classical physics: the theory did not always provide a precise trajectory of the electron, but rather indicated the probabilities of the possible outcomes of a measurement.

Schrödinger's equation thus became a fundamental tool for understanding the structure of atoms, the formation of molecules and chemical bonds, the properties of solids, and the behavior of materials. Much of modern technology—from semiconductors to lasers, from transistors to spectroscopic diagnostics—depends, directly or indirectly, on that conceptual revolution launched in Zurich in 1926.

Although Schrödinger never had the temperament of a technician confined to the serenity of the laboratory, he never felt entirely at ease with the philosophical conclusions that many drew from the s of quantum physics. He had helped build one of its pillars, but he was not willing to turn uncertainty into a convenient intellectual surrender. He was wary of overly hasty formulas, overly confident analogies, and interpretations that seemed to invite physicists to abandon the idea of a comprehensible reality.

In 1933, he received the Nobel Prize in Physics, sharing it with Paul Dirac. That same year, Adolf Hitler came to power in Germany. Schrödinger was teaching in Berlin at the time, where he had been awarded the prestigious chair formerly held by Max Planck. But the new regime made German intellectual life increasingly stifling. Although he was not Jewish, Schrödinger refused to conform to the political and cultural atmosphere of Nazism; he left Germany in the summer of 1933.

Thus began a long period of upheaval: Oxford, Graz, Rome, exile, and the search for a permanent home in a Europe that was becoming increasingly dangerous. Schrödinger's biography is not that of a scientist serenely protected by universities and academies. It is the biography of a man torn apart by the fractures of his century.

The most significant destination during this phase was Dublin. In 1940, Irish Prime Minister Éamon de Valera—who had studied mathematics and retained a genuine interest in scientific research—founded the *"Dublin Institute for Advanced Studies.*" The institute was conceived as a place where top-level scholars could work without the obligation of daily teaching and without being overwhelmed by the urgencies of war. Schrödinger became director of *the "School of Theoretical* Physics" and remained in Dublin until 1956.

It is a cultural scene less well-known than those of Copenhagen, Berlin, or Princeton, but one worth imagining. While Europe was at war, in a neutral, “ ” Ireland, Schrödinger gave lectures and reflected on the nature of life. He was not seeking an escape from physics; he was seeking to push its boundaries to a new frontier.

In February 1943, he delivered a series of lectures at *Trinity* College in Dublin titled **What Is Life?**—which would become, the following year, a short but extraordinarily influential book. The text did not offer a complete biological theory. It could not have done so: the structure of DNA would not be identified by James Watson and Francis Crick until ten years later, in 1953. But Schrödinger posed a question that many biologists would take seriously: How can an organism preserve and transmit order in a universe where, spontaneously, things tend toward disorder?

He spoke of an “aperiodic crystal”—that is, a material structure capable of storing information without repeating itself monotonously like a normal crystal. He had not discovered DNA, and it would be incorrect to portray him as a prophet who already knew of its double helix. But his insight helped shape the imagination of a generation. Francis Crick recalled that **What Is Life?** was one of the books that sparked his interest in biological problems; James Watson read it as a young man and was deeply impressed by it. Sometimes, in the history of science, an idea is valuable not because it immediately provides the solution, but because it shows where to look.

This is one reason why Schrödinger appears to be an unconventional figure: he did not meekly respect the boundaries between disciplines. After contributing to atomic physics, he turned his attention to biology. And when biology seemed to him insufficient to fully address the problem of human experience, he did not hesitate to turn to philosophy.

Of course, here we must avoid a very common temptation: that of portraying Schrödinger as a mystic disguised as a scientist, or of using his name to claim that quantum physics has proven the doctrines of the East. This is not the case. The Upanishads do not contain quantum theory; "Advaita Vedānta" does not anticipate laboratory experiments; no finding in physics justifies the claim that "consciousness creates reality" in the vague and consolatory sense that this phrase often takes on.

But it would be equally wrong to deny that Schrödinger was deeply interested in Indian thought. He read and meditated on texts from the Vedantic tradition, including through the translations and studies of the German philosopher and Sanskrit scholar Paul Deussen. He was particularly drawn to one question: Does it really make sense to imagine consciousness as a multitude of small, isolated entities, each enclosed within its own body and its own history? Or does individual experience point to a deeper continuity, one that is difficult to express using conventional categories?

In **Mind and* Matter*, published in 1958, Schrödinger writes: *"Subject and object are one and the same."* This is a statement that calls for caution. It is not a theorem of quantum mechanics. It does not derive from the equation that bears his name. It is a philosophical reflection on the fact that we observe the world from within the world: we cannot completely step outside of reality to describe it as if we were not part of it.

Physics, at this point, can tell us many precise things: how particles behave, what probabilities we can calculate, and which experiments confirm or refute a theory. But when we ask ourselves what it means, for a conscious being, to be part of the cosmos it seeks to understand, we enter different territory. No less serious, but different. Here, science is not replaced by philosophy; it is complemented by questions that science, with its tools, cannot resolve on its own.

Schrödinger was not always consistent, nor was he personally exemplary. His private life was complex and at times troubled; his philosophical ideas do not enjoy unanimous agreement; some of his statements on consciousness remain debatable. We do not need to turn him into a saint of reason or a retrospective spiritual master. His importance lies precisely in the restlessness he embodies.

A complex personality.

This book aims to be objective and to avoid both hagiography and hasty condemnation. Schrödinger was an extraordinary figure, but not a straightforward one: a great physicist capable of questioning the very foundations of his own discipline, a thinker drawn to the spiritual unity of reality, and at the same time a man whose private conduct raises real ethical issues. His "unease" lies precisely in the fact that he offers no reconciled synthesis.

Schrödinger was not always consistent, nor was he personally exemplary. His private life was complex and, in many respects, problematic. He had been married since 1920 to Annemarie Bertel, who remained his wife for the rest of his life, but their marriage was not exclusive by conventional standards. Schrödinger had numerous extramarital affairs; some were lived openly, or at least tolerated, within an unusual and fragile family dynamic.

The best-known case is his affair with Hilde March, the wife of physicist Arthur March. A daughter, Ruth Georgie, was born from that relationship in 1934. Annemarie continued to be a part of Schrödinger's life, and during the years of exile and later in Dublin, the web of family relationships remained complex. This is not enough to reduce a person to his romantic history, but it prevents us from portraying him as an ascetic of knowledge, detached from human conflicts.

There is also documentary evidence—primarily diaries and correspondence—of relationships with very young women. Some episodes, viewed through today's lens but also in light of the disparities in age, prestige, and power that characterized them, are ethically troubling. It is not necessary to turn these facts into gossip or use them to deny the value of his work; however, it would be equally wrong to ignore them. A thinker may have formulated groundbreaking ideas and, at the same time, engaged in personal behavior that does not deserve celebration.

This ambivalence calls for a fundamental distinction: understanding an author does not mean absolving him; acknowledging his dark side does not mean erasing his scientific contribution.

Even in physics, Schrödinger was less "Schrödingerian" than the popular image suggests. The equation that bears his name is one of the pillars of quantum mechanics; however, he never readily accepted the probabilistic interpretation that became dominant.

When Max Born proposed in 1926 that the square of the wave function should be interpreted as the probability of finding a particle at a certain position, Schrödinger greeted the idea with considerable reservations. He would have preferred that the wave function describe something physically real and continuous—an actual distribution of matter or charge—rather than a mere statistical prediction of measurement outcomes. But this interpretation faced difficulties: a wave packet associated with a particle tends to spread out in space, whereas an observed particle appears localized.

The famous cat, therefore, was not invented to celebrate the idea that an animal could be "truly" both alive and dead at the same time. It was a critique. In 1935, Schrödinger sought to demonstrate how paradoxical it seemed to extend quantum language indefinitely from the atomic world to everyday objects. If an atom can exist in a superposition of states and its decay triggers a lethal mechanism, then—following certain formulations of the theory to the letter—

a cat, too, should exist in a superposition of life and death until it is observed. For Schrödinger, this was a sign of conceptual difficulty, not a fanciful description of reality.

Yet he did not offer a complete and widely accepted alternative solution. He saw the measurement problem with extraordinary clarity, but he did not solve it. This is one of his great legacies: having articulated a theoretical dilemma that physics, even today, addresses through competing interpretations—from Copenhagen to the many-worlds interpretation, from collapse theories to hidden variables.

The unity of consciousness: a philosophical insight, not a conclusion of physics.

In the final decades of his life, Schrödinger turned increasingly explicitly to philosophy, biology, and the traditions of India, particularly the Upaniṣads and "Advaita Vedānta." In texts such as *"My View of the* World" (published posthumously in English in 1964), he argued that the multiplicity of individual consciousnesses might, at a deeper level, be merely apparent.

His most provocative formulation asserts, in essence, that consciousness is "singular": there are not many separate consciousnesses, but a single reality of feeling that manifests itself through different individuals. This thesis is close to the non-dualistic doctrine of the ātman and the Brahman: the inner self and the ultimate foundation of the cosmos are not two distinct realities.

This insight may be philosophically fruitful, but it is not a proven consequence of quantum mechanics. Schrödinger sometimes drew a parallel between the two realms—the crisis of the rigid distinction between observer and observed in quantum physics and the non-duality of the Vedic traditions—but the comparison remains analogical and metaphysical. It does not constitute scientific proof that "consciousness creates reality," nor that the human subject causes the collapse of the wave function.

Here we must avoid a very common oversimplification: Schrödinger did not scientifically prove "Advaita Vedānta" and did not establish a quantum theory of consciousness in the contemporary sense of the term. Rather, he expressed a personal, erudite, and intense philosophical vision, born of the encounter between the conceptual crisis of modern physics, the European philosophical tradition—Schopenhauer had a significant influence on him—and his reading of Indian texts.

The scientific community's objections to this position are serious. If there is only one consciousness, how can we explain the radical difference between individual experiences? How can we account for a person's concrete suffering, the autonomy of other people, personal memory, and moral responsibility? And above all: what experimental procedure could confirm or refute this hypothesis? Today, as we speak of a "field of consciousness," the debate is more relevant than ever.

The book **What Is Life?** also highlights the tensions in Schrödinger's thinking. The book was profoundly influential: it suggested that heredity must depend on a stable molecular structure capable of containing information—the famous "aperiodic crystal." This insight helped guide a generation of young researchers, including Francis Crick and James Watson, toward the problem of the structure of genetic material.

But the book should not be mythologized. Schrödinger did not discover DNA nor did he deduce its double helix; he relied in part on data and models that were incomplete at the time. His merit was not in offering a definitive biological theory, but rather in posing a powerful question: how can living matter preserve order and transmit information without defying the laws of physics?

Even the famous expression "feeding on negative entropy" is evocative but imprecise according to contemporary thermodynamic terminology. Organisms do not literally "eat" negative entropy: they maintain their organization locally by exchanging energy and matter with the environment and exporting entropy. Nevertheless, behind that imperfect formulation lay a genuine and generative question: how is order maintained in living things?

Thus, Schrödinger was neither a saint of reason nor a spiritual master to be retrospectively canonized. He was a man of immense intelligence, but also of contradictions. His private life, marked by multiple relationships and at times by behavior that is difficult to defend today, cannot be set aside in the name of scientific greatness. Nor should his philosophical ideas on the unity of consciousness be mistaken for findings proven by quantum physics: they belong to a personal metaphysical inquiry, nourished by Schopenhauer and the Upanishads, not to an experimentally verified theory.

Even Schrödinger's cat stems from a sense of unease, not from certainty. Schrödinger had formulated the fundamental equation of wave mechanics, but he never fully reconciled himself with the probabilistic " " interpretation of the theory. His thought experiment was not intended to proclaim that a cat was truly both alive and dead at the same time; rather, it sought to demonstrate just how problematic it was to transfer quantum formalism unmediated into the world of ordinary experience.

Erwin Schrödinger's importance perhaps lies precisely here: not in having offered a perfect, coherent, and reassuring doctrine, but in having fully engaged with the fractures of modernity. Between wave and particle, chance and necessity, life and matter, individual and totality, Schrödinger never ceased to seek a language capable of holding together what seemed irremediably separate. His questions continue to challenge us.

In a century that was learning to divide knowledge into ever-narrower fields, Schrödinger continued to ask inconvenient questions. How can matter become alive? What distinguishes an organism from an aggregate of atoms? What place does consciousness have within the framework of nature? And why does the human being, who studies the world with such precision, still struggle to understand what it means to exist?

II. Erwin Schrödinger: The Philosopher-Physicist.

"Consciousness cannot be explained in physical terms. Consciousness is absolutely fundamental."
Erwin Schrödinger, *Mind and Matter* (1958)

"Multiplicity is only apparent; in truth, there is but a single consciousness."
Erwin Schrödinger, *What Is Life?* (1944)

"The scientific picture of the real world around me is very lacking. It gives us a great deal of factual information, arranges all our experience into a magnificently coherent order, but remains frighteningly silent on everything that truly matters to us."
Erwin Schrödinger, *Nature and the Greeks and Science and Humanism* (1954)

Erwin Schrödinger is best known as one of the founders of quantum mechanics. Yet to limit him to the equation that bears his name would be to miss the most remarkable aspect of his legacy. For him, physics was never an activity separate from the ultimate questions: what is life, what relationship exists between mind and nature, why do we perceive ourselves as distinct individuals, and whether this distinction truly exhausts what we are.

In this chapter, therefore, we will encounter neither a spiritual master turned scientist nor a scientist who found in physics proof of ancient wisdom. We will encounter a modern thinker—rigorous and restless—convinced that equations are indispensable but cannot, on their own, tell us everything we wish to understand. His research moves along the boundary: where what physics can state with precision ends and what human beings can honestly question begins.

Erwin Schrödinger belongs to that rare breed of scientists who have never regarded their discipline as a closed room. For him, physics was not simply the art of predicting the behavior of matter. It was a way of looking at the world, and at the same time an invitation to recognize how much broader the world remains than the descriptions we are able to give of it.

This mindset made him, from the very beginning, something of an outsider. Not because he disdained precision—on the contrary, he helped forge one of the most effective mathematical tools in contemporary science—but because he did not accept that precision equaled completeness. An equation can describe with extreme accuracy how a physical system will evolve. Yet it does not tell us what the color we see is, what the pain we feel is, what value we attribute to a life, or what the word “I” means.

It is important to grasp this distinction, because it avoids two mirror-image errors. The first consists in demanding answers from physics that it is unable to provide: for example, a scientific proof of the soul, of free will, or of God. The second consists in believing that, since physics does not answer these questions, they are meaningless. Schrödinger accepted neither conclusion. Rather, he sought to keep open the space between verifiable knowledge and lived experience.

Born in Vienna on August 12, 1887, Schrödinger grew up in a city that, before the Great War, was an extraordinary cultural laboratory. Vienna in the late nineteenth and early twentieth centuries was not merely the capital of the Habsburg Empire: it was the home of Gustav Mahler and Arnold Schönberg, of Sigmund Freud and Arthur Schnitzler, of Karl Kraus and Ludwig Wittgenstein. In those streets, among the cafés, universities, concert halls, and literary journals, science was not yet entirely separated from philosophy, art, and Europe's moral crisis.

Erwin's father, Rudolf Schrödinger, was an entrepreneur with interests in science and botany; his mother, Georgine Emilia Brenda, was the daughter of Alexander Bauer, a chemist of English origin who taught at *the "Technische* Hochschule" in Vienna. English was also spoken in the Schrödinger household. This bilingual and cosmopolitan upbringing left its mark: Erwin would not merely be an Austrian physicist, but a European in the broadest and most vulnerable sense of the word.

When he enrolled at the University of Vienna in 1906, he found an environment still influenced by the legacy of Ludwig Boltzmann, who had committed suicide that same year. Boltzmann had been one of the great pioneers of statistical physics, but also a combative thinker, involved in profound disputes over the very meaning of scientific theories . Schrödinger was not his direct student; instead, he studied under Franz Serafin Exner, an experimental physicist and a leading figure in the Vienna School. Nevertheless, Boltzmann's intellectual presence remained strong: physics appeared not merely as a collection of results, but as a human endeavor that compels us to rethink the most obvious images of reality.

Perhaps this is where we can identify Schrödinger's defining trait. It was not enough for him to know that a theory worked. He wanted to understand which worldview made that functioning possible. Not in the sense of seeking an elegant metaphor to add to the calculations, but in the stricter sense of asking whether, behind the effectiveness of the formulas, there remained an intelligible vision of nature.

Quantum mechanics emerged in a climate of enormous creativity and equally enormous unease. Between 1925 and 1927, in various parts of Europe—Göttingen, Munich, Zurich, Copenhagen, Cambridge—a small group of physicists developed theoretical tools capable of describing the atomic world with a precision never before achieved. Werner Heisenberg developed matrix mechanics; Max Born clarified its probabilistic application; Paul Dirac provided a formulation of great power and conciseness. Schrödinger, working in Zurich, introduced wave mechanics in 1926. We cannot overlook the Italian contribution, that of Enrico Fermi. In 1926, almost simultaneously and independently of Paul Dirac, Fermi formulated the statistics applicable to particles subject to the Pauli exclusion principle, such as electrons. It is now known as Fermi-Dirac statistics; the particles that follow it are called fermions.

His work, published under the title *"On the Quantization of the Monoatomic Ideal Gas,"* described the behavior of a gas of identical particles in which no two particles can exist in the same quantum state. It was a decisive result for understanding the electronic structure of atoms, the properties of metals, electron degeneracy, and, later, astrophysical phenomena such as white dwarfs and neutron stars.

Thus, while Italy did not produce, in those years, an author of a general formulation of quantum mechanics comparable to Heisenberg's matrix mechanics or Schrödinger's wave equation, Fermi made a foundational contribution to its statistical framework and its physical applications. In the years that followed, Fermi's Roman school—with Rasetti, Amaldi, Segrè, Pontecorvo, and, most notably, Ettore Majorana—would become one of the leading schools of theoretical and nuclear physics in Europe.

Quantum theory changed the history of physics. But now is not the time to trace its technical development: we will do so in the next section of this chapter. Here, it suffices to note one revealing fact. Schrödinger did not see the new theory merely as a solution to a laboratory problem. He saw in it a sign that the familiar categories—object, motion, matter, observation—were becoming less straightforward than they had seemed in classical physics.

In the tradition pioneered by Isaac Newton, the world had often been imagined as an immense architecture of bodies arranged in space, governed by precise laws that were, at least in principle, predictable. This framework was not naive: it had yielded immense results. But twentieth-century physics had begun to undermine it. First, Einstein's theory of relativity had altered our concepts of space and time; then, quantum physics had shown that, in the realm of the infinitely small, it was not always possible to attribute to a particle distinct properties independent of how we observe it.

Here, physics has the final word. It does not authorize us to say that "everything is an illusion," nor that the mind invents the world as it pleases. It states something more sober and, perhaps, more unsettling: the descriptions that work for objects of everyday life do not translate without modification to the atomic scale. Nature is not bound to coincide with the images that are familiar to us.

Schrödinger made a decisive contribution to this revolution, but he never ceased to recognize its problematic aspect. His dissent was not directed against the experiments, which confirmed quantum predictions in an increasingly convincing manner. It was directed at the abandonment of understanding that, in his view, risked accompanying the practical success of the theory. In other words: Can we really make accurate calculations without knowing what we are actually saying about the world? To many physicists, the question might have seemed secondary. If a theory allows for correct predictions, perhaps that is all we need to expect. Schrödinger, however, felt that this position, while legitimate on an operational level, left an essential part of the intellect unsatisfied. In a 1950 lecture, later included in the collection *_Science and_ Humanism*, he argued that the task of science is not merely to produce usable knowledge, but to contribute to a coherent understanding of humanity's place in nature. The point was not to reject specialization—which is inevitable in a science that has become vast—but not to mistake a part for the whole.

One of his famous quotes is: *"The scientific picture of the world around me is sorely lacking. It provides me with a great deal of factual information; it organizes all our experience in a magnificently coherent way; but it is terribly silent about everything that is truly close to our hearts, about what really matters to us."*

Schrödinger is not accusing science of having failed. On the contrary, he acknowledges that it organizes facts into a "magnificently coherent" order. But he observes that the language of physics, precisely because it is so powerful, makes a choice: it describes the world in quantitative, measurable, and shareable terms. To do so, it sets aside a person's immediate experience: the taste of wine, the red of a sunset, fear, joy, remorse, anticipation. This is not an accidental flaw in science. It is the price of its precision. Temperature can be measured with great accuracy, but the sensation of cold is not a number. The brain's electrical activity can be recorded, but the intimate quality of a memory does not appear on an instrument's graph. We can describe the frequencies of light, but no formula captures the experience of blue as it appears to the person who sees it. Saying this does not mean pitting a supposed "soul" against matter, as if they belonged to two universes that have nothing to say to one another. Schrödinger was not seeking refuge from science. Rather, he was seeking a less truncated vision, in which human beings would not be treated as alien spectators facing an entirely impersonal nature.

It is here that his reflection becomes more nuanced and moves further away from clichés. Schrödinger took a long-standing interest in the philosophical traditions of India, particularly the Upanishads and "Advaita Vedānta." He did not do so to adorn physics with spiritual exoticism. He was searching for words and images capable of addressing a question that, in his view, European culture tended to leave unresolved: why does each of us feel like a separate center of experience, even though we are all part of the same nature?

His engagement with Indian thought was also shaped by European scholars such as Paul Deussen, a German philosopher and Sanskritist, and , author of important works on the Upanishads and Vedānta. Schrödinger did not possess the philological expertise required to read Sanskrit; his was a philosophical engagement, mediated by translations and the intellectual climate of his time. This limitation must be clearly stated. But it must also be said that his interest was not superficial: in Vedantic ideas, he found a radical way to challenge the image of the self as a completely isolated entity. When he writes that the multiplicity of minds might be merely apparent, Schrödinger is not formulating a thesis proven by quantum mechanics. This is not a matter of experimental physics. Here, a philosophical proposal—influenced by the Vedantic tradition—comes into play, one that can be accepted, discussed, or rejected. Its importance lies perhaps above all in the question it inevitably raises: Is the separation between "me" and "the world"—between the knowing subject and the known reality—truly as absolute as it seems to us in everyday experience?

In Western culture, this question has ancient roots. We can trace it back at least to Baruch Spinoza, who in the seventeenth century conceived of mind and body not as two separate substances, but as two ways of understanding a single reality. And we can recall, in a very different context, William James, the American philosopher and psychologist who, in the late nineteenth century, described consciousness not as a series of static objects, but as a flow. Schrödinger is neither Spinoza nor James. However, he shares with them a rejection of an overly simplistic division between the observer and the observed.

In him, however, this unease had a particular intensity. It did not stem from a desire to construct a flawless philosophical system. It arose from the impression that modern science, precisely as it explained nature ever more thoroughly, had made it more difficult to answer a fundamental question: Who is the one who knows?

In 1944, in Dublin, Schrödinger published **What Is Life?**, a book destined to have a lasting influence on the nascent field of molecular biology. But even when he was addressing the material foundations of heredity, his perspective did not stop at the mechanism itself. Life appeared to him as a physical problem, certainly, but also as the point at which organized matter comes to question itself.

This idea must be handled with caution. It does not mean that an atom “thinks” or that the universe has hidden intentions. These are claims that science does not support. It means, quite simply, that human beings are nature: the brain that formulates a theory, the hands that build a tool, and the eyes that read a page all belong to the same world that the theory attempts to describe. Nature, through one of its living forms, has become capable of asking questions about itself.

Schrödinger died in Vienna on January 4, 1961. He had returned to his hometown after years spent in Ireland and a life marked by wars, exiles, political upheavals, and scientific revolutions. His famous equation is engraved on his tomb in the Alpbach cemetery in the Austrian Tyrol. It is an eloquent image: a few mathematical symbols—sober and almost abstract—signifying one of the great turning points in modern thought. But for Schrödinger, that equation was never a tombstone set over the mystery. It was a gateway. Schrödinger does not offer us a final answer on existence, consciousness, or the unity of reality. He offers us something less reassuring and more fruitful: the courage not to consider the quest over once the equation has been found.

When matter learned to make waves.

Before his name became inextricably linked to an imaginary cat, Erwin Schrödinger was, above all, the man who gave quantum physics a new language—a language made of waves. It was neither a minor gesture nor a mere technical refinement. When, in 1926, Schrödinger published his famous works on wave mechanics, quantum theory had already existed for about a quarter of a century, but it was still a restless, fragmented construct—rich in results but lacking in shared imagery. It explained certain atomic phenomena with extraordinary effectiveness, but it seemed to speak a language too distant from ordinary experience. There were sudden energy jumps, quantum numbers, selection rules; and, since 1925, the matrices of Werner Heisenberg, Max Born, and Pascual Jordan: powerful mathematical objects, yet not very intuitive even for many physicists.

Schrödinger had the impression that, behind that new physics, there might exist a more continuous description, closer to the grand visions of classical science. He did not want to go back, nor did he want to deny the novelty of quantum theory. Rather, he sought to understand whether the apparent discontinuity of the atomic world could be expressed through something more familiar: the propagation of a wave.

It is important to clarify: we are still in the realm of physics here, not metaphysics. The idea was not that "everything is vibration"—a suggestive but overly vague formula to convey anything precise. The idea was more rigorous and, precisely for this reason, more surprising:

perhaps material particles, such as electrons, possess real and measurable wave-like properties.

The hypothesis had been put forward a few years earlier by Louis de Broglie, a young French aristocrat who had initially studied history and then physics. In his doctoral dissertation, defended at the Sorbonne in Paris in 1924, de Broglie took a bold step. If light—considered for centuries to be a wave—could also manifest itself as a collection of quanta, or photons, why couldn't matter exhibit the reverse behavior? Why couldn't an electron, which we imagine as a particle, also have a wavelength?

De Broglie formulated the idea in clear terms: *"One must assume that every portion of energy endowed with its own mass is associated with a periodic phenomenon.*" The sentence appears in his 1924 work *"Recherches sur la théorie des* quanta." Behind the somewhat solemn language of the time lies a simple insight: matter and energy do not belong to two separate worlds; both can manifest, depending on the circumstances, as particles or waves.

The thesis initially met with skepticism. Paul Langevin, who served as its advisor, nevertheless recognized its originality and forwarded it to Albert Einstein. Einstein, who had made a decisive contribution to the concept of light quanta, immediately recognized its significance. It was one of those insights that, at first, seem almost too simple to be true: if nature compelled light to be both a wave and a particle, perhaps there was no reason to deny matter the same duality.

Schrödinger took up that hypothesis and gave it a complete mathematical form. In Zurich, where he had been teaching at the university since 1921, he worked in an intellectually vibrant environment, permeated by a question that at the time was not merely technical: what picture of the world emerges from the new physics? The

Swiss city, suspended between the precision of its academic life and the proximity of the Alps, had also been the place where Einstein had developed some decisive aspects of general relativity. This is not merely a geographical detail. Early 20th-century physics did not develop solely in laboratories: it flourished in universities, in correspondence among scientists, in café discussions, and during mountain retreats, in a Europe striving to regain its footing after the devastation of the Great War.

During a stay in Arosa, in the canton of Graubünden, between late 1925 and early 1926, Schrödinger worked intensively on the theory that would change his destiny. The starting point was an old analogy, developed nearly a century earlier by the Irish mathematician William Rowan Hamilton. Hamilton had shown that the motion of bodies and the propagation of light share a profoundly similar mathematical structure. In optical geometry, for example, a ray of light is not an isolated object: it is a line indicating the direction of propagation of a wavefront, just as the crest of a sea wave suggests a direction without fully capturing the phenomenon.

Schrödinger turned this perspective on its head: if the trajectories of classical mechanics resemble the rays of optics, perhaps the motion of particles is merely an approximation—one that holds when the waves associated with matter have wavelengths that are minuscule compared to the dimensions of the objects. A planet can be treated as a point traveling along an orbit; an electron in an atom, however, requires a different description. Not as a tiny particle circling the nucleus along a well-defined trajectory, but as a wave function.

Schrödinger stated this with almost programmatic clarity in the first article of the series titled _“Quantisierung als Eigenwertproblem,”_ published in _the *Annalen der_ Physik* in 1926: _“The task consists in replacing the_ *problem of the motion of material points with the problem of wave propagation.*" This sentence encapsulates a shift of enormous significance. It was no longer a matter of asking where exactly the electron was and which path it was following. One had to ask what waveform was possible in a given physical situation.

Here lies one of the most fruitful images of wave mechanics. Consider a violin string. It cannot vibrate in just any way: if it is fixed at both ends, it admits only certain stable forms of oscillation. The simplest produces a fundamental note; other, more complex forms generate harmonics. There is no indifferent continuum of possible notes: there are modes permitted by the structure of the string.

Similarly, according to Schrödinger, an electron bound to an atomic nucleus cannot assume just any energy state. Only certain "standing waves"—forms of vibration compatible with the conditions imposed by the atom—are possible. It is here that quantization—which until then had often been introduced as a rule seemingly foreign to classical physics—gains a more concrete representation. The discrete energy states of the atom are no longer, at least in this picture, steps placed there arbitrarily: they are the possible harmonies of a wave system.

The decisive case was the hydrogen atom, the simplest of all atoms: a nucleus consisting of a single proton and a single electron. In 1913, Niels Bohr had succeeded in explaining certain lines in the hydrogen spectrum by assuming that the electron could occupy only certain orbits. It had been a success, but it left many questions unanswered. Why those particular orbits? And why did the electron, when moving from one to another, emit or absorb light at precise frequencies?

Schrödinger's theory yielded the same energy levels without having to imagine the electron as a tiny planet orbiting the Sun. The old planetary model of the atom—so effective in textbooks and so difficult to abandon—began then to reveal its limitations. The atom was not a miniature solar system. It was something more foreign to our intuition, yet also more orderly: a system in which certain waveforms are allowed and others are not.

Its equation—now called the Schrödinger equation—quickly became a fundamental tool. There's no need to delve into its symbols here. It suffices to grasp the essential idea: given a physical situation, the equation allows us to calculate how a wave function—usually denoted by the Greek letter ψ, "psi"—evolves over time. This function is not a material wave in the common sense of the term; it is not made of water, air, or matter oscillating in space. It is a mathematical description of the system's state, capable of producing extremely accurate predictions.

This distinction is crucial. Saying that the electron "is a wave" may be useful as an initial mental image, but it becomes misleading if taken literally. In experiments, the electron is revealed as a localized event: it leaves a trace on a screen, ionizes an atom, or strikes a detector. Yet, before detection, its behavior is governed by a wave function that can extend through space and produce interference, just as light does.

In 1927, experiments by Clinton Davisson and Lester Germer in the United States, and those by George Paget Thomson in Great Britain, provided spectacular confirmation of de Broglie's hypothesis: electron beams produced diffraction patterns as they passed through crystals. It was a quintessentially wave-like phenomenon. The son of J. J. Thomson—the scientist who had discovered the electron as a particle in 1897—thus helped demonstrate that the electron also possessed a wave-like nature. The history of physics sometimes favors such family ironies.

Wave mechanics did not remain a competing theory to Heisenberg's matrix mechanics for long. Schrödinger demonstrated, as early as 1926, that the two formulations were mathematically equivalent: they said the same things using different languages. Matrix mechanics was more abstract and algebraic; wave mechanics offered concepts closer to the problems of optics, vibrations, and differential equations. It was not a victory of one over the other, but the discovery that the new physics could be viewed from two perspectives.

Paul Dirac, one of the great architects of quantum theory, would write in 1929 a phrase that has since become famous: "*The fundamental physical laws necessary for the mathematical theory of most of physics and all of chemistry are therefore fully known.*" It was a bold statement, perhaps even excessive when viewed through today's lens. But it captured the spirit of an era: with quantum mechanics, a framework had finally been established capable of explaining the stability of atoms, chemical bonds, and the properties of matter.

The very word "chemistry" deserves a moment's consideration. Schrödinger's theory did not concern only invisible particles in laboratories. It concerned the shape of molecules, the way electrons are distributed around nuclei, and the very possibility that stable bonds exist between atoms. In other words, it helped clarify the physical language through which matter becomes a mineral, a molecule, a cell, a pigment, a drug, a living body. Of course, it did not explain life as a whole—it would be a huge mistake to think so—but it provided an indispensable part of the grammar by which matter organizes itself.

One delicate question remained, however: what exactly does the wave function represent? Schrödinger, at least initially, hoped that it might describe a sort of continuous physical distribution of the electron. But this interpretation ran into difficulties. In that same year, 1926, Max Born proposed a decisive shift: the wave function does not indicate where the particle is, as if it were a diffuse cloud of matter; rather, it allows us to calculate the probability of finding it in a given location.

Born formulated this new idea in a sentence of crystal-clear radicalism: *"The motion of particles follows the laws of probability, but probability itself propagates according to the law of causality."* The first part of the statement broke with a long-standing aspiration of physics: to know the present conditions exactly in order to predict the future exactly. The second part, however, served as a reminder that the quantum world was by no means a realm of pure chance. The wave function evolves according to precise rules; what appears probabilistic is the individual result of the measurement.

Schrödinger was never entirely satisfied with this point. But his unease belongs to the next stage of our story. For now, what matters is to see what he had achieved: he had transformed a fragmented theory into an extraordinarily powerful tool for calculation and thought; he had given quantum mechanics a dynamic, continuous image, capable of uniting the language of waves with the discrete structure of the atom.

His equation does not invite us to an easy poeticization of nature. On the contrary, it compels us to exercise discipline in our imagination. It tells us that what we call a "particle" cannot always be thought of as a small, compact entity traveling along a predetermined path. In the infinitely small, reality presents itself first as a set of possibilities ordered by a wave- l law, and only then—when we encounter an experimental result—as a determined event.

It is from this rift—both scientific and human—that Schrödinger's unease with the dominant interpretation of quantum theory will arise. But before delving into that dissent, we must recognize the nature of his gift: not an escape from the equations, but a deeper way of engaging with them.

When Equations Are No Longer Enough.

Wave mechanics had been one of Erwin Schrödinger's great triumphs. In 1926, his equation had given quantum theory a form of extraordinary effectiveness: it made it possible to calculate the states of the atom, to understand the spectra of light emitted by chemical elements, and to approach that invisible structure upon which matter, molecules, and chemical bonds depend. Yet the very man who had contributed decisively to building that language never felt fully at home in the world it seemed to describe.

It was not a case of nostalgia for classical physics. Schrödinger was not a man incapable of accepting innovation. He had just introduced one of the most fruitful formulations of the new physics and had demonstrated, in that very same year of 1926, that his wave mechanics and Werner Heisenberg's matrix mechanics were two mathematically equivalent ways of expressing the same theory. He did not dispute the results. He disputed the haste with which many colleagues seemed willing to transform those results into a philosophy of resignation.

For him, the problem began with a seemingly simple question: what does the wave function actually describe? Schrödinger's equation allows us to calculate the evolution of a function denoted by the Greek letter ψ, psi. Initially, Schrödinger hoped that this function might represent something physically extended: not a tiny invisible particle, but a continuous distribution in space. It was an imperfect image, but it had the advantage of preserving a certain continuity: something exists, transforms, and evolves according to a law.

In July 1926, however, Max Born proposed a different interpretation. The wave function, he argued, does not describe matter dispersed throughout space; it describes the probabilities of the possible outcomes of a measurement. If ψ is large in a certain region, it does not mean that there is "more electron" there, just as there would be more fog at a certain point in a valley. It means that the probability of detecting the electron is greater there, if an appropriate measurement is made.

It was an enormously successful proposal. It accurately explained the statistical nature of atomic experiments and was consistent with what was observed in laboratories. But it opened up a rift that was hard to ignore. Before the measurement, the theory seems to describe not a result that is already present but simply unknown, but rather a multitude of possibilities. After the measurement, however, we find a precise fact: here, not there; this value, not that one; one detector on, the others off.

How does this transition occur? The theory, in its standard formulation, contains two very different processes. On the one hand, Schrödinger's equation causes the wave function to evolve in a continuous and deterministic manner. On the other hand, when a measurement is made, a single outcome emerges, which the theory can predict only in terms of probability. In textbooks, this transition has often been called the "collapse of the wave function": a convenient term, but one that does not solve the puzzle. What collapses? For what reason? At what exact moment? And above all: is measurement a physical process like any other, or does it introduce a special break into nature?

Schrödinger refused to accept that such questions should be dismissed as idle. A theory can be impeccable in its calculations and yet leave us uncertain about what it is actually saying. This is no minor flaw, especially if the theory in question claims to describe the fundamental building blocks of matter.

In the 1920s, the most influential interpretation of the new physics developed around Niels Bohr in Copenhagen and Werner Heisenberg: the so-called "Copenhagen School." It would, however, be too simplistic to speak of a "Copenhagen doctrine" as if Bohr, Heisenberg, Born, and the others had signed a single manifesto. Their positions did not coincide in every detail. They did, however, share a radical caution: quantum physics does not allow us to describe the atomic world using the familiar images of everyday life. We cannot unreservedly attribute a definite position and velocity to an electron before measuring them; we cannot imagine it as a tiny planet traveling along a hidden trajectory.

Bohr called this situation "complementarity." Under certain experimental conditions, a quantum object exhibits properties we call particle-like; under others, it exhibits wave-like properties. These are not two incomplete snapshots of the same classical reality that eludes us: they are different aspects, both necessary, but not always representable together using ordinary categories.

Heisenberg, reflecting many years later on that historic turning point, coined a phrase *that* has since become famous: *"What we observe is not nature in itself, but nature as it is revealed by the way we ask questions.*" The phrase appears in his book *Physics and Philosophy*, published in 1958, and deserves to be read carefully. It does not mean, as is sometimes claimed, that the human mind invents the world or that each person can produce their own reality through thought. It means that, in quantum physics, the experimental setup is not an external detail: it determines which property can be revealed and in what form it can be recorded.

This caution had genuine power. It prevented us from projecting overly reassuring images onto the atomic world. But to Schrödinger, it seemed that, when used too hastily, it could become an alibi: the theory works; therefore, let's stop asking what lies behind the results.

It is here that his dissent takes on an almost moral tone, as well as a scientific one. For Schrödinger, physics was not a catalog of successful predictions. It was an attempt to construct an intelligible picture of nature. Of course, a provisional, correctable picture, perhaps destined to be superseded. But still a picture. If science gave up asking what existed and how things could hold together, it risked being reduced to a sophisticated technique for anticipating instrument readings.

His objection was not a childish desire to return to a completely predictable universe. Schrödinger knew full well that nature has no obligation to be intuitive. Nor did he believe that a better interpretation of quantum mechanics could restore the orderly world of Newtonian physics. His unease concerned something more specific: he found it unacceptable to treat as definitive an obscure separation between the physical evolution of the system and the act of measurement.

In 1927, at the Fifth Solvay Conference in Brussels—one of the most intense gatherings in the history of twentieth-century science—this tension was already evident. The group photograph taken on that occasion features, among others, Albert Einstein, Marie Curie, Max Planck, Paul Dirac, Louis de Broglie, Heisenberg, Bohr, and Schrödinger himself. It is an almost surreal gallery of intellects, but not of people in agreement. Quantum theory was by then powerful enough to assert itself; it was not clear enough to pacify those who had created it.

Einstein, despite having contributed more than anyone else to the birth of quantum theory, shared some of Schrödinger's concerns. His famous phrase—*"God does not play* dice"—has been repeated ad nauseam and is often misinterpreted. Einstein was not asking for a consolatory physics, nor was he invoking a deity to correct the equations. He was expressing the conviction that probability might indicate an incomplete description, not the final word on reality.

Bohr replied with an equally famous quip: *"Albert, stop telling God what to do.*" Beyond the anecdote, the clash between the two was not a dispute between reason and irrationality, nor between old and new. It was a disagreement over the very purpose of physics. Einstein and Schrödinger feared that the dominant interpretation was giving up too soon on the search for a reality independent of our acts of observation. Bohr believed that the call for a more classical picture of reality was not a matter of greater depth, but a question posed in the wrong language.

In 1935, the issue became even more acute. In May, Albert Einstein, Boris Podolsky, and Nathan Rosen published an article in the journal **Physical* Review* that was destined to go down in history by the initials of its authors: the EPR paradox. Their goal was not to deny the correctness of quantum calculations. They wanted to show that, if certain consequences of the theory were taken seriously, two particles that had interacted could maintain correlations such that the quantum description would appear incomplete—or, at the very least, profoundly at odds with our ordinary notion of spatial separation. In other words, two particles could remain in contact— —as if they were a single particle, even if separated by astronomical distances.

Schrödinger read that article carefully and exchanged an intense correspondence with Einstein that summer. It was then that he introduced the German term *"Verschränkung,"* now translated as "entanglement" or "quantum entanglement." Schrödinger realized that this entanglement was not a side issue, but *"the defining feature of quantum* mechanics"—the very thing that truly distinguished it from classical physics.

Two systems can interact and then move apart. Quantum theory can describe them no longer as two fully separable realities, each with its own independent state, but as a single system distributed throughout space. Today, experiments have rigorously confirmed the correlations predicted by the theory. Despite the doubts raised by the EPR paradox, entanglement is a confirmed reality. However, precisely because entanglement is real in the experimental sense of the term, Schrödinger's question has not become any less important: what does this tell us about the nature of separation, of the object, of physical individuality?

That same year, Schrödinger published, in three parts in the German journal **Die* Naturwissenschaften*, an essay titled *"The Current Situation in Quantum Mechanics."* It is here that the most famous cat in the history of science makes its appearance.

It is worth noting that the cat was not invented to demonstrate that animals can be both alive and dead at the same time, nor to promote the idea that human consciousness creates the world. On the contrary: it was a critique. Schrödinger deliberately constructed an absurd example to illustrate the discomfort that arises when quantum language is transferred—without clarification—into the world of large objects, living bodies, and everyday experience.

The experiment has already been described previously, but it is worth recalling its main aspects. Let's imagine a closed box. Inside are a cat, a tiny amount of radioactive material, a Geiger counter, a mechanism with a hammer, and a vial of poison. If a radioactive atom decays within a certain time, the counter registers it, the hammer breaks the vial, and the cat dies. If the atom does not decay, nothing happens and the cat remains alive.

Before opening the box, quantum theory assigns the radioactive nucleus a superposition of the two possibilities: decayed and not decayed. But if the decay is causally linked to the animal's fate, then the system as a whole also appears to be described by a superposition: a live cat and a dead cat.

Schrödinger wrote ironically that, considering quantum theory, one can *"construct quite ridiculous cases*." The point was precisely this: if the interpretation of the wave function is taken as a direct snapshot of the state of affairs, must we really say that a living being is, prior to observation, in a mixed state of life and death?

The immediate answer should be a cautious one. It is incorrect to claim that quantum mechanics literally asserts that an ordinary cat is "half-alive and half-dead" until a human being lifts the lid. That is a popular simplification, often turned into a metaphor for consciousness or indecision, but it does not exhaust either physics or the philosophical problem. Measurement does not necessarily require a conscious mind: in laboratories, an automatic detector records an event even in the absence of any human observer.

And yet the paradox does not disappear with this correction. If it is the detector that performs the measurement, then we must ask ourselves where the process begins and where it ends. Is the Geiger counter a quantum object? What about the hammer? The vial? The air in the box? And the cat, with its billions upon billions of particles and its irreducible status as a living being?

Contemporary physics has developed important tools to address some of these questions. The theory of decoherence, developed primarily since the 1970s, explains how a system's ceaseless interaction with its environment renders quantum superposition extremely fragile and unobservable on a macroscopic scale. A cat is not an isolated atom: it breathes, exchanges heat, emits and absorbs light, and continuously interacts with its environment. All of this contributes to the emergence, in our ordinary world, of stable and well-defined outcomes.

But decoherence, however essential, does not by itself resolve every issue. It explains very well why we do not see cats in superposition; it leaves open—according to various interpretations—the question of why a single outcome appears to have actually occurred. That is why Schrödinger's cat continues to live on in scientific culture: not as a magical image, but as a persistently concrete question about the relationship between equations, phenomena, and the world of experience.

Schrödinger's unease thus stemmed from a demanding fidelity to physics, not from a rejection of physics. He did not want to replace experiments with spiritual intuitions, nor to introduce mysteries where calculation had already brought clarity. Rather, he insisted that the practical success of a theory not be confused with a full understanding of its meaning.

There is something in this request that still concerns us today. We are accustomed to living surrounded by technologies born of quantum mechanics—semiconductors, lasers, magnetic resonance imaging, atomic clocks—and we might be tempted to conclude that a theory capable of producing such results has now settled every debate. Schrödinger reminds us of the opposite: a theory can be incredibly precise and still hold a question within it. This is not a weakness of knowledge. It is its most vital aspect.

Science Confronted with the Question of Being.

"The scientific picture of the real world around me is very incomplete. It gives us a great deal of concrete information, organizes all our experience in a magnificently coherent way, but remains terribly silent on everything that is truly close to our hearts, on what really matters to us."
Erwin Schrödinger, *Nature and the Greeks and Science and Humanism*, 1954.

"Something unknown is doing something we do not know: that, in essence, is what our theory consists of."
Arthur S. Eddington, *The Nature of the Physical World*, 1928.

"Science cannot solve the ultimate mystery of nature. And this is because, in the final analysis, we ourselves are part of the mystery we are trying to solve."
Max Planck, *Where Is Science Going?*, 1932.

Some scientists leave behind a theory; others leave, along with a theory, a new way of looking at the world. Erwin Schrödinger belongs to this second, rarer, and perhaps more unsettling category. His wave equation, formulated in Zurich in 1926, is one of the cornerstones of modern physics: it made it possible to describe the quantum states of electrons, atoms, and molecules, paving the way for a vast swath of contemporary science, technology, and . Yet, for Schrödinger, the success of that equation by no means settled the matter.

A theory can work with extraordinary precision and, at the same time, leave open the simplest and most difficult question: what is it really telling us about the world in which we live?

It is not a matter of demanding from physics what physics cannot provide. That would be a mistake mirroring that of those who, when faced with the great questions of humanity, dismiss science as a cold and partial endeavor. Physics is not a book of consolation; it is not meant to dictate how to live, nor to decide what is right to love, fear, or hope for. Its task is more precise and, precisely for this reason, immensely powerful: to construct testable descriptions of natural phenomena, to relate different experiences, to formulate laws, and to predict outcomes.

But every description, to be effective, makes selections. And what it selects inevitably leaves something out of the picture. When a physicist describes light as electromagnetic radiation, he is not denying the red of a sunset or the blue that appears among the clouds after a thunderstorm. He is simply speaking in a different language. That language measures wavelengths, frequencies, and energies; it does not capture the nostalgia, joy, or sense of openness that a landscape can evoke in those who contemplate it. These dimensions are not false simply because they do not appear in an equation. They are dimensions of lived experience that the equation, by its very nature, was not designed to encompass.

Alfred North Whitehead, an English mathematician and philosopher, expressed this point in deliberately stark terms. The nature of physics, he wrote in 1925, is *"without sound, without smell, without color."* Of course, Whitehead did not mean that the real world is truly gray and silent. He meant that the physical portrait of the world—if we take it in isol —has had to set aside the sound we hear, the color we see, and the scent we perceive. Not because these are illusions in the trivial sense of the term, but because they belong to the way the world unfolds for a living, conscious being.

Schrödinger felt this distance particularly keenly. He was no enemy of scientific precision: it would be absurd to think that of one of the architects of quantum mechanics. He was, however, wary of the idea that precision could be transformed into a total philosophy, capable of explaining everything simply because it explains so much.

In 1954, in *"Nature and the Greeks and Science and Humanism,*" Schrödinger articulated one of his clearest reflections. The scientific picture of the world, he observed, is "quite incomplete": it organizes facts magnificently, but remains silent on what truly moves us. It is a statement that deserves to be understood without haste. It is not an attack on science; it is a call for moderation. Physics possesses an extraordinary ability to describe how things behave. But human existence is not limited to the behavior of things.

We can know a great deal about the chemical composition of tears and still understand nothing about grief. We can measure the electrical activity of a person's brain while listening to Bach and still not have grasped the full meaning of that listening experience. We can describe the biological processes of birth with ever-greater precision and yet stand before it with the same sense of wonder that the ancients felt.

We need to be rigorous here. It does not follow at all that science is incapable of studying the mind, emotions, or life. Neuroscience, biology, experimental psychology, and medicine have much to say, and they say it with often decisive results. The point is another: no external, description can completely replace the experience from within. A map is indispensable for traversing a territory, but it does not contain the wind, the fatigue in one's legs, or the smell of the earth after the rain.

Schrödinger did not seek to bridge this gap by resorting to facile spiritualism. He was too well-versed in the discipline of physics to confuse an intuition with a proof. However, he also did not accept that, in order to be scientific, one must refrain from asking questions about the relationship between matter, life, mind, and the totality of reality. For him, science was an adventure of human reason; and human reason, when it is alive, never ceases to question the place it itself occupies in the cosmos.

This tension accompanied him throughout his life. Born in Vienna in 1887, Schrödinger grew up in a city that, before World War I, was one of Europe's great intellectual laboratories. It was the Vienna of Gustav Klimt and Arnold Schönberg, of Sigmund Freud and Ludwig Wittgenstein; a city where music, medicine, art, philosophy, and physics could still come together in the same conversations, without blurring into one another. Schrödinger studied physics at the University of Vienna, but he also received a strong classical education: he knew Greek and Latin, read ancient philosophy, loved poetry, and cultivated a serious interest in the history of ideas throughout his life.

This is no mere ornamental detail. For him, returning to the Greeks did not mean fleeing from modernity. It meant remembering that science was born from a broader question: What is the world made of? What order exists in nature? What relationship unites human beings with what they observe?

The Ionian thinkers—Thales, Anaximander, Anaximenes—sought a common principle underlying the diversity of things. Their answers may seem naive today: water, the àpeiron, air. Yet the intellectual gest that animated them remains decisive. They were not content with arbitrary narratives; they wanted to find coherence in nature. Modern physics is immensely more precise, more mathematical, and more capable of prediction. But it retains, in a new form, that same confidence: the world is not pure disorder, and the mind can grasp at least some of its structures.

Schrödinger felt that this search for unity must not come to a halt in the face of the fragmentation caused by specialized fields of knowledge. Specialization is necessary: without it, we would have neither atomic physics nor modern medicine nor precision astronomy. But it can have a side effect: it can make us forget that the many disciplines speak, each in its own way, of the same world. The world of the chemist, the biologist, the physicist, the historian, and the poet is not composed of five separate realities. They are five perspectives, five tools, five different degrees of approach to a reality that no one possesses entirely.

This is why Schrödinger could take an interest in biology without abandoning physics. When, during his years in Dublin, he delivered the lectures in 1943 that would become the book *_What Is Life?_*, he did not claim to offer a definitive theory of life. Rather, he posed a question: How can matter organize itself in such a way as to preserve and transmit an order as delicate as that of living organisms? Molecular genetics had not yet emerged in the form we know today; the double-helix structure of DNA would be identified by James Watson and Francis Crick only ten years later, in 1953. Yet Schrödinger's little book helped many young researchers realize that life was not a realm foreign to physics, but a frontier capable of demanding new ideas.

It is significant that this question was posed in Dublin during the war years. Schrödinger had arrived there in 1939, after leaving continental Europe, which had been thrown into turmoil by the Nazi expansion. Ireland, which had remained neutral during the conflict, was then home to the newly established *"Dublin Institute for Advanced Studies,"* founded on the initiative of Éamon de Valera, an Irish politician and mathematician by training. In that relatively secluded environment, Schrödinger headed the School of Theoretical Physics. His lectures on life took place at *"Trinity College Dublin"*: a place steeped in history, yet open at that moment to one of the most modern questions that could be posed.

The question was modern not because it was new in an absolute sense, but because it shifted the boundaries between fields of knowledge. What distinguishes a living being from an inanimate structure? How does an organism maintain its form over time, even as matter and energy constantly pass through it? How does biological order resist the general tendency of physical systems to disperse differences?

Schrödinger tackled these problems using the categories available at the time, some of which would later be corrected, refined, or superseded by subsequent research. But his merit lies not only in his answers. It lies in having shown that a scientific question can simultaneously be a question about how we conceive of ourselves. If life is matter organized in an extraordinarily complex way, this statement does not automatically reduce it to something insignificant. Instead, it forces us to ask ourselves what we mean by "matter," by "organization," and by "meaning."

These are words we often use as if they were self-evident. They are not.

Classical physics had offered a picture of the world composed of localized bodies, forces, and trajectories. In principle, if all the initial conditions were known, one could calculate the future. Quantum mechanics did not simply replace one set of formulas with another: it called into question some of the intuitive images that underpinned that picture. At the atomic level, we cannot always attribute definite properties to a particle independently of how we observe it; predictions take on a probabilistic form; everyday language encounters real limits.

Arthur Eddington, an English astronomer and physicist, summed up the problem with a memorable touch of irony: *"Something unknown is doing we-don't-know-what."* This is neither an admission of total ignorance nor a call to abandon research. Eddington had made a decisive contribution to the experimental verification of Einstein's general relativity by observing the deflection of starlight during the solar eclipse of May 29, 1919. He knew full well the limits of scientific knowledge. Precisely for this reason, he knew that, behind the solidity of the laws, there remains a question about the ultimate meaning of what we call reality.

Schrödinger shared this sense of restraint. Mystery is not a gap to be hastily filled with solemn words. Nor is it proof of some spiritual doctrine. It is the provisional name for what resists our simplifications. When physics states that a quantum system is described by a wave function, it is making a technical assertion, verifiable through experimental consequences. When we ask ourselves whether this description implies that the world is made up of possibilities, relationships, events, or something we do not yet know how to name, we are already in the realm of interpretation. And when we ask ourselves what place consciousness has in a universe that becomes self-aware through beings like us, we enter a further realm: no longer in the inevitable conclusions of physics, but in philosophical and existential reflection.

Distinguishing these levels does not impoverish the discourse; it makes it freer and more honest. It avoids two opposing temptations. The first is reductionism: believing that, since physics successfully describes certain aspects of reality, it has already said everything that can be said. The second is facile mysticism: using words like "quantum," "energy," or "observer" to justify whatever belief we wish to hold.

Schrödinger did not belong to either of these tendencies. He sought a broader vision without turning science into religion and without separating spiritual reflection from the responsibility of thought. His interest in Indian traditions—and in particular in the thought of the Upanishads and "Advaita Vedānta"—must be understood in this light. Not as a purported demonstration that quantum physics "confirms" ancient doctrines—which would be historically and scientifically incorrect—but rather as the recognition of a possible harmony: both modern physics and certain great philosophical traditions can lead us to be wary of the overly simplistic image of a completely isolated self standing before an alien world.

But this is only a threshold. Before crossing it, it is worth lingering a moment longer with Schrödinger the physicist. His most valuable lesson is not a ready-made answer about the unity of the cosmos. It is an attitude: taking the equations seriously without granting them the final word on what it means to exist.

Max Planck, one of the fathers of quantum theory, put it clearly: we are part of the mystery we study. We do not observe the world from a balcony outside the universe. We are bodies among bodies, living beings among living beings, minds capable of formulating theories and, at the same time, of questioning the value of that very ability.

Perhaps this is where science truly becomes a worldview: not when it claims to solve every mystery, but when it teaches us to see more clearly just how close the mystery is.

III. The Cat That Didn't Want to Die Twice.

"The ψ function of the entire system would express this by stating that it contains the live cat mixed or spread out, in equal parts, with the dead cat."
Erwin Schrödinger, "The Present Situation in Quantum Mechanics" (1935).

"The word 'measurement' has had such a detrimental effect on discussions of quantum mechanics that, in my opinion, it should now be banished entirely from quantum mechanics."
John S. Bell, "Against 'Measurement,'" 1990

In 1935, Erwin Schrödinger imagined a cat locked in a steel box along with a device as simple as it is sinister: a small radioactive source, a Geiger counter, and a mechanism capable of breaking a vial of hydrogen cyanide. If, within a certain time interval, an atom decays, the counter registers the event, the mechanism is triggered, and the poison is released. The cat dies. If the atom does not decay, the cat remains alive.

It was not an experimental proposal, and certainly not a cruel game. No cat was placed in a box in Schrödinger's laboratory. It was a thought experiment: a philosophical tool designed to test the words we use to describe the world.

In the quantum description, before a measurement is made, a microscopic system can exist in a superposition of possibilities. It is not simply like a coin still face-down on a table, where we do not know which side is facing up. In the case of the coin, heads or tails is already decided: all that's missing is our knowledge of it. In the quantum case, at least according to the most common interpretation of the theory, the situation cannot be described so easily. The formalism assigns the system a combination of possible outcomes, with well-defined probabilities.

Schrödinger took this language to its most jarring conclusions. If the atom can be described as both decayed and undecayed, and if the cat's fate depends on the atom, then should the cat also be described as both alive and dead at the same time? This is where the paradox stops us in our tracks. Not because it forces us to believe in an animal suspended between life and death, but because it reveals how difficult it is to move seamlessly from the world of electrons and atoms to that of bodies, living beings, and human experiences.

This chapter explores that divide. We will see that the cat was not invented to prove that quantum physics is absurd, nor to legitimize any fantasy about consciousness "creating" reality. Schrödinger, on the contrary, wanted to highlight a serious problem: if the theory works extraordinarily well in predicting experimental outcomes, how should we understand the transition from quantum possibilities to a concrete fact? When, and in what sense, does an event become an event?

The question does not concern only the human observer. Is a Geiger counter already an observer? What about photographic film? A molecule colliding with another molecule? The entire environment? The word "observation" can be misleading, because it seems to imply a conscious gaze. In much of contemporary physics, however, what matters is the physical interaction: one system leaves traces in another system, and those traces effectively become irreversible.

Yet the cat retains its agency because it is neither the needle of an instrument nor a speck of dust. It is a living being. It breathes, perceives, reacts, and has an organized and vulnerable body. Its presence does not automatically introduce a new law of physics; biology does not suspend quantum mechanics. But it forces our imagination to grapple with what formulas alone cannot capture: the difference between describing a process and understanding what it means, for a living being, to be alive or dead.

The cat, the box, and the blind spot of physics.

When we talk about Schrödinger's cat, it's best to clear up a very common misunderstanding right away. The paradox does not claim that, in everyday life, every cat locked in a room is literally 50 percent alive and 50 percent dead until someone opens the door. Nor does it say that the human mind possesses a magical power to create reality with a glance.

It says something more sober—and perhaps more unsettling: quantum theory, rigorously applied at the microscopic level, provides us with a description of possibilities; but we experience facts. We see a detector that clicks or that does not click. We find an image on a photographic plate or we do not. And, in the extreme case imagined by Schrödinger, we encounter a living animal or a dead one.

The thought experiment appeared in 1935, in a series of articles published in the German journal **Naturwissenschaften**. It was a pivotal year. A few months earlier, Albert Einstein, Boris Podolsky, and Nathan Rosen had published the famous article now known by the acronym EPR, in which they argued that quantum theory, however effective, might be incomplete. Schrödinger entered the debate with a partial sympathy for Einstein's critique, but with a voice entirely his own.

He was not hostile to quantum mechanics: he had made decisive contributions to its foundation. The equation that bears his name remains one of the pillars of modern physics to this day. Precisely for this reason, his challenge carried weight. Schrödinger was not mocking a theory he did not understand. He was questioning a theory he had helped to build.

The crux of the difficulty is superposition. In the language of quantum physics, an isolated radioactive atom can be described by a state that includes both possibilities: decaying and not decaying. We must not immediately translate this statement into crude imagery. It does not mean that the atom is a tiny ball split in half between two fates, like a character in a fantasy story. It means that the theory does not assign the system a single one of the two outcomes prior to a certain interaction that makes the outcome measurable.

But the device in the box links the atom's fate to a chain of increasingly larger events: the decay, the Geiger counter, the mechanism, the vial, the poison, the animal's body. If we apply the formalism without pausing, the initial uncertainty seems to propagate along the entire chain. And here is the deliberately provocative formula: the cat would be both alive and dead.

Schrödinger aims to bring to light a point that risks remaining hidden behind technical jargon. If we take literally the idea that superposition remains valid at every scale, we would have to accept descriptions of the world that do not correspond in the least to our experience. We never see a cat that is half-alive and half-dead. We never encounter tables that are simultaneously in two different rooms, nor bells that ring and do not ring in exactly the same sense.

The first reaction might be: so quantum mechanics is false. But that would be a hasty conclusion. Its predictions have been confirmed with unprecedented precision: in the structure of matter, in semiconductors, in lasers, in magnetic resonances, and in the electronics found in our homes. The problem isn't that the theory doesn't work. The problem is understanding exactly what the theory is telling us.

Here we must distinguish between different levels. Physics tells us that microscopic objects can exhibit interference and superposition; it tells us that quantum probabilities are calculable and verifiable. Interpretation attempts to clarify what this description means: Are the alternatives real at the same time? Is there a physical- l process—still unknown—that selects an outcome? Does each outcome occur in a different branch of reality? Finally, philosophical reflection asks what we mean when we say that something "is" in a certain way, and what place conscious experience has in the overall picture.

Confusing these three levels leads to two mirror-image errors. The first is hasty reductionism: "Physics has proven that consciousness counts for nothing." It has not proven this. The second is facile mysticism: "Physics has proven that the mind creates the world." It has not proven this either.

The decisive—and at the same time the most deceptive—word is "measurement." In a laboratory, measurement means relating a system to an apparatus capable of producing a record: a flash, a number, a trace, an electrical signal. However, as soon as we try to pinpoint exactly where the quantum system ends and where the apparatus measuring it begins, we find ourselves faced with a line that is difficult to draw.

The Irish physicist John Bell, one of the great thinkers of the second half of the 20th century, was wary of that very word. Not because physicists shouldn't measure, of course, but because "measurement" seems to introduce into the theory a concept taken from everyday life without fully clarifying its meaning. We measure using instruments made of atoms. Those instruments interact with other atoms. Where does the transition occur between quantum behavior and observable phenomena?

An important answer comes from decoherence, a concept developed primarily in the 1970s and 1980s, thanks in part to the work of the German physicist Heinz-Dieter Zeh and, later, Wojciech Zurek. In simple terms, a large, hot object that is continuously exposed to its environment does not remain isolated. It interacts with photons, air molecules, vibrations, and thermal radiation. These interactions rapidly disperse the delicate relationships necessary for a superposition to remain observable.

An atom can be carefully shielded within an experimental setup. A cat, on the other hand, is a universe of cells, chemical exchanges, heat, respiration, and nerve impulses; moreover, it is immersed in its environment. Even if it were enclosed in a box, it would never be a simple, perfectly isolated system. Decoherence explains why we do not observe cats in clearly superimposed quantum states. The environment erases—or rather, renders practically irrecoverable—the interference between the various possibilities.

This is a fundamental achievement, but it does not settle the entire issue. Decoherence explains very well why certain alternatives cease to interfere with one another and why the macroscopic world appears stable. But it leaves one question open: if the overall description still contains different possibilities, why do we experience precisely this outcome, and not another?

This is at the heart of what is called the measurement problem. It is not a marginal gap for specialists. It is the point at which physics, without losing its rigor, encounters the limits of its usual frameworks.

Some, such as Eugene Wigner, have suggested that consciousness might play a role in the transition to a definite outcome. This is a historically important proposal, because it brings to light a difficulty that many preferred to ignore. However, it has not become an experimental conclusion in physics. Other physicists believe that interaction with the environment is sufficient; still others propose that the wave function undergoes, under certain conditions, a real physical collapse; others accept the idea of a s many branches, in which the possible outcomes are realized separately.

None of these approaches offers an easy answer. And this is perhaps the most honest lesson of Schrödinger's cat: it does not ask us to hastily choose a metaphysical framework, but to recognize that the experience of the world is harder to describe than we imagine.

Living beings enter into the paradox not as an exception to the laws of nature, but as a reminder of their scope and their descriptive limits. Physics can describe the chemical reactions occurring within an organism, the flow of energy, the structure of molecules, and the electrical activity of nerve cells. But when we say "this cat is alive," we are also pointing to a dynamic unity: a being that sustains itself, that perceives its environment, that is subject to pain, hunger, fear, and death.

This is not about adding a soul as an explanation for what biology has not yet understood. That would be a shortcut. It is about not confusing a description—however refined—with the full reality of what is being described. An accurate map of Florence does not contain the smell of wet stone after the rain, the sound of the Arno, or the exertion of those climbing toward San Miniato al Monte. And yet, the map remains invaluable. Its limitation is not a flaw: it is the price of its precision.

The same is true of physics. Its formulas are not impoverished because they do not recount the experience of a living animal. They serve a different purpose: they identify regularities, relationships, probabilities, and structures. But we may ask ourselves whether the world, as we inhabit it, is exhausted by those regularities.

The cat in the box continues to gaze at us from this very boundary. It offers us no reassuring solution. It does, however, prevent us from falling into two forms of laziness: that of those who wish to reduce the mystery to a technical term, and that of those who use the mystery to evade precision.

Perhaps its most enduring lesson is this: when reality appears paradoxical to us, we need not immediately choose between surrender and fantasy. We can linger in the question. We can distinguish between what physics asserts, what interpretations propose, and what, beyond both, still remains to be thought.

Alice is not in Wonderland: she is locked in a box.

In 1935, Erwin Schrödinger was forty-eight years old, teaching at Oxford, and already one of the leading figures in physics of his time. Two years earlier, he had received the Nobel Prize for formulating wave mechanics; his equations had provided the new physics with a language of extraordinary power. Yet he, of all people—one of the architects of quantum theory—felt that something was amiss in the way that theory was presented and, perhaps even more so, in the way it was interpreted.

The cat stems from this unease. It is not a witty gimmick devised to make quantum physics memorable; nor is it a foreshadowing of a certain kind of sensationalist popular science, which loves to transform science into a repertoire of parlor paradoxes. It is a thought experiment constructed with extreme polemical precision. Schrödinger intended to show what happens if we take the language of quantum superposition literally, without any caution, and apply it from the realm of atoms to that of everyday objects, and finally to that of a living being.

The case is well known. A cat locked in a box; the presence of the radioactive atom is not random. In the quantum world, prior to measurement, we generally cannot represent certain processes as if they already possessed one and only one definite outcome, simply hidden from us. Instead, the theory assigns a state— —that combines different possibilities. In the scenario imagined by Schrödinger, the atom is described by a combination of two alternatives: decayed and undecayed; that is, it is "in a superposition of states."

This word, "superposition," has met with a curious fate. It has become familiar, almost elegant; it circulates in novels, in advertisements, and in conversations about creativity. But in its physical sense, it does not at all indicate a vague indecision, nor the simple fact that we do not know how things stand.

If a coin is spinning on a table and we cannot yet see whether it will land heads or tails, the coin is not, for that reason, simultaneously heads and tails. It already has one side facing up, even if it is hidden from us. Our ignorance does not alter what is happening on the table. Quantum superposition is something else entirely: it is the way the theory describes a system in which those alternatives cannot be treated—before the interaction that makes one of them visible—as simple, already established facts that are merely unknown to us.

Here we must proceed with rigor, but also with caution. Saying that a particle is "in two states at once" may be useful as an initial mental image, but it risks misleading us. The words "together," "here," "there," "before," and "after" belong to our experience of solid objects and well-defined situations. Quantum mathematics does not necessarily authorize us to imagine a tiny object simultaneously playing two roles in the same way that an actor might stand on a stage wearing an ambiguous costume. Rather, it offers us a structure of possibilities, in which the outcomes are not yet separated as they will be when the system leaves a readable trace.

The difference is crucial. Superposition is not an irrational fantasy introduced by physics to shock common sense. It is an element of the formalism that has verifiable consequences and that, in the microscopic world, manifests itself in an extremely concrete way. Interference experiments are one of its clearest manifestations. When beams of electrons, photons, or even molecules pass through carefully designed devices, they produce distributions that do not correspond to the behavior of tiny projectiles endowed, from the outset, with a predetermined trajectory.

Quantum physics does not, therefore, ask us to believe in the absurd. It asks, more modestly, that we abandon certain overly simplistic mental images. It asks us to admit that, on very small scales, the question "What exactly is this particle doing?" may not have the kind of answer that common sense expects.

But the cat enters the scene precisely to highlight a limitation. If the radioactive atom is described by a superposition, and if the counter is linked to the atom, and the mechanism to the counter's reading, and the vial to the mechanism, and the cat's fate to the vial, then it seems that the entire chain must inherit that same description. No longer just an atom that has decayed and one that has not, but a device that is activated and one that is not; a vial that is broken and one that is intact; and finally, in Schrödinger's deliberately scandalous formulation, a cat that is both alive and dead.

This is where formalism encounters common sense—not to be easily refuted by it, but to be forced to explain itself. Common sense protests: a cat cannot be both alive and dead. Not because death and life are two incompatible notions, nor because humans are not intelligent enough to tolerate a bit of strangeness. It protests because life and death are not minute, reversible properties, like the direction of polarization or the value of a quantity measured in a laboratory. They are comprehensive conditions. They involve an organism, a biological h , a temperature, an endless array of chemical processes, an environment, and a body capable of reacting—or incapable of doing so. A living cat is not an enlarged atom; a dead cat is not simply the "other value" of a variable.

Schrödinger chooses a cat precisely because the cat is not neutral. A rock would have had less of an effect; a light bulb, whether on or off, would have left the reader more at ease. An animal, on the other hand, brings the problem of life into the box. It brings breath, metabolism, vulnerability, and the possibility of pain; it also brings, at least implicitly, a form of experience. The provocation does not lie in imagining a creature magically suspended between two fates. It lies in asking whether it is truly legitimate to transfer, without residue, the language developed for the microcosm to an entire organism.

There is an important detail that is often overlooked. Schrödinger was not saying, "Quantum mechanics proves that cats are truly both alive and dead at the same time." He was arguing almost the opposite: if a certain interpretation of the theory forces us to speak seriously of a live-dead cat, then that interpretation leaves something essential unresolved in the transition from the microscopic to the world we inhabit.

The paradox, therefore, does not mock quantum physics from the outside. It is a critique born at its very heart. Schrödinger knew the power of his equations all too well to dismiss them in the name of everyday intuition. But he also knew that a formalism, however effective, is not automatically a complete picture of reality. It can precisely calculate what will happen in an experiment and yet leave open the question of what, exactly, it is describing.

At this point, it is worth distinguishing between different levels. On the level of physics, superposition is a necessary and component of the theory and is confirmed by experiments. On the level of interpretation, the question of how to understand the relationship between that mathematical description and the definite events we observe remains open. On the human and philosophical level, the cat forces us to acknowledge something simple: our most immediate words—"exists," "happens," "is alive," "is dead"—suddenly become problematic when we claim to use them in the same way at every scale of nature.

This is not an invitation to suspend judgment on everything, nor a license to claim that "anything is possible." On the contrary: it is an exercise in precision. Quantum physics does not claim that a person can simultaneously have left and remained at home, or that every contradiction is true in some poetic universe. Quantum possibilities obey rigorous laws; they are expressed by strict mathematical relations, not by a vague celebration of ambiguity.

And yet, the cat paradox leaves a fruitful wound in our image of the world. It reminds us that reality is not obliged to present itself immediately in the form that the human mind finds most comfortable. For centuries, classical physics had accustomed us to imagining a universe composed of things endowed with definite properties, located in specific places, and subject to identifiable causes. The quantum world does not entirely erase this image, but it limits its scope. It warns us that, prior to certain encounters and certain observations, nature can be better described as a web of possibilities than as an inventory of ready-made facts.

Schrödinger's box thus remains closed even within us. Not because it contains an exotic enigma, but because it challenges a deep-seated habit of ours: the assumption that reality must always coincide with the way we see it once it has already become an everyday experience . The cat does not teach us that reason fails; it teaches us that reason, when it is truly true to itself, knows how to recognize the point at which it must question its own images. And that question, once raised, no longer concerns only an atom, a device, and an animal locked in a box. It concerns the uncertain boundary between what theory can express and what we call, with apparent simplicity, a fact.

The problem of measurement.

The cat paradox almost immediately leads to a question that, when phrased in everyday language, seems simple: when does a possibility become a fact?

This is not a marginal question, reserved for specialists in quantum interpretations. It is the point at which the most effective theory ever constructed to describe matter encounters a difficulty of language and imagery. Quantum mechanics can predict the probabilities of an experiment's outcomes with extraordinary precision. But when we try to ask it what happens in the transition from the set of possibilities to a specific result, the answer does not appear quite as unambiguous.

Let's take an example less dramatic than Schrödinger's cat. In a laboratory, a single photon is sent toward a device capable of detecting its passage. At the end of the process, a small bright spot appears on the screen; or a detector emits a click; or a computer records a number. Something has happened. Not a nebulous cloud of possibilities, not an unresolved statement: a precise, legible, shareable signal.

The problem arises before that click. According to the quantum description, the photon can be represented as a distribution of possibilities: not as a tiny particle that necessarily follows a single, predetermined path, but as something that evolves according to rules different from those of our mechanical intuition. As long as the system remains isolated, the theory tracks its evolution with continu ity and rigor. Then the measuring apparatus intervenes: the detector, the screen, the photographic plate, the electronic circuit. And we obtain a result. But why that particular result?

It's not enough to say: because we looked. In physics, "observing" does not primarily mean opening one's eyes. A detector has no consciousness, makes no judgments, and is not surprised by what it records. It can remain on in an empty room overnight and record events without anyone witnessing them. A particle leaves a trail in a cloud chamber; a photon strikes a sensitive material; an atom collides with a sensor and alters the state of a device. When the researcher arrives the next morning and consults the data, the recording has already taken place.

This is the first misunderstanding to clear up: quantum physics, in its standard form, does not claim that the human mind materially creates the world through an act of attention. The idea is seductive, and it has fueled countless spiritualistic oversimplifications. But the fact that consciousness is a great mystery does not give us the right to use it as a magic word to fill every gap in physics.

The measurement problem is more sober and, precisely for this reason, more profound. It concerns the boundary—if such a boundary exists—between a system described in terms of quantum possibilities and an event that appears to be definite. In 1927, in Brussels, during the Fifth Solvay Conference, this question was at the center of one of the most famous intellectual debates of the twentieth century. Niels Bohr and Albert Einstein were not arguing over the practical results of the theory: Einstein was well aware of its effectiveness. They were debating what the theory allowed us to say about the world.

Bohr, a central figure of the Copenhagen school, insisted on a point that may seem modest but is by no me : to communicate the results of an experiment, we must describe the instruments and the outcomes using ordinary language—the kind that distinguishes one position from another, a signal from silence, a stationary pointer from a moving one. The measuring device must, in some sense, belong to the world of established facts. We cannot do physics, Bohr suggested, without being able to say what was actually recorded.

Einstein, however, sensed the cost of this position. If quantum theory works perfectly for atoms, electrons, and photons, why should we introduce, at some point, a separate domain—that of "classical" instruments—where things possess a definite outcome? Where exactly does the boundary lie? In the detector? In the circuit that amplifies the signal? In the computer that stores the data? In the experimenter's retina? In his consciousness?

The question is not a trick question. If we keep moving the boundary a little further ahead, it seems as though we can avoid it; but it keeps reappearing. It is as if a line drawn in the sand recedes every time we try to reach it.

The mathematician and physicist John von Neumann clearly saw this difficulty. In 1932, in his book **The Mathematical Foundations of Quantum* Mechanics*, he showed that we can apply quantum laws not only to the particle under study but also to the apparatus that measures it. And then, at least in principle, a second device that records the first, and so on. This forms what is often called the "von Neumann chain": system, instrument, recording, observer.

Every link in the chain can be included in the physical description. But if everything is included, where does the single result we find at the end come from?

This is physics as it is most commonly formulated. Before the measurement, the system is described by a structure of possibilities. At the moment of measurement, the theory introduces an additional rule: among the possible outcomes, only one manifests itself, with a calculable probability. This transformation is often called the "collapse" of the wave function. The term is effective, but it can evoke the image of a small building collapsing inside the atom. Nothing of the sort happens. It is a name given to the transition—predicted by the theory—from a description containing multiple possible outcomes to a single, determined result.

The difficulty lies in the fact that the two modes of description seem very different. On the one hand, we have a continuous, precise evolution governed by equations. On the other, a single event, unpredictable in its details, even if predictable in terms of its probabilities. The theory tells us how often, upon repeating the experiment many times, a certain result will appear. It does not tell us why, in that single instance, the detector recorded precisely that result and not another.

It is important not to interpret this situation as a failure of science. In laboratory practice, quantum mechanics works impressively well. It underlies semiconductors, lasers, magnetic resonance imaging, and an essential part of contemporary technologies. No engineer needs to solve the problem of measurement before building a transistor or operating a sensor.

Yet effectiveness does not erase the question of meaning. We can know how to use a map without having fully resolved what the territory is. Science, when it is honest, does not hide this distinction.

In the 1970s, the German physicist Heinz-Dieter Zeh and, later, the Polish-American Wojciech Zurek developed a line of research that is now fundamental: decoherence. The general idea is that no macroscopic object is truly isolated. A speck of dust, a drop of water, an organism, or a laboratory- strument constantly exchanges energy and information with its surroundings: light, heat, air molecules, radiation, and vibrations.

These interactions make truly quantum behavior extremely fragile when we move on to large and complex objects. The possibilities that, in a well-isolated microscopic system, might still interfere with one another become practically indistinguishable from the environment. The everyday world thus acquires the stability we are familiar with: tables that stay in one place, needles that point to a specific position, readings that do not waver before our eyes between incompatible results.

Decoherence is a decisive theoretical breakthrough because it shows how the environment contributes to the formation of the stable and well-defined nature of things. It is not necessary to wait for a human gaze for the physical world to leave traces of itself. Reality, so to speak, continually "writes" itself into its environment.

But it would be unwise to attribute more to decoherence than it actually offers. It explains very well why, in everyday life, we do not observe the strange interference patterns typical of isolated quantum systems. It explains why certain possibilities cease to be observable as mutually influencing alternatives. However, for many physicists and philosophers of science, one question remains open: if the overall description preserves all possibilities, why do we encounter only a single outcome?

This is not a subtle point for specialists. It is the question contained in the click of the detector. Why this click, here, now?

John Bell, the Northern Irish physicist who in 1964 formulated the famous theorem destined to change the debate on the foundations of quantum mechanics, possessed a dry sense of irony and little tolerance for verbal explanations. In a 1981 essay, *"Quantum Mechanics for* Cosmologists," he posed the question provocatively. The wave function of the universe, he asked, would it have had to wait billions of years for the emergence of a living creature to "jump" to a result? And, he added mockingly, was it perhaps necessary to wait for someone "with a Ph.D."?

Bell's objection hits a specific target: if we link the result of the measurement to the observer's consciousness, we risk making the universe strangely dependent on the belated arrival of human beings. Stars emitted light for billions of years before there were eyes capable of seeing it. Nuclear reactions took place in the hearts of stars, minerals formed, and planets took shape long before any earthly mind could utter the word "measurement."

Consciousness cannot be casually introduced like a cosmic switch. This does not mean it is irrelevant. It means that its role, if it exists, must be considered more seriously. Consciousness is what transforms a result into experience, knowledge, memory, and shared language. A detector can produce a signal; only a conscious being can ask what that signal means, what context it fits into, and what image of reality it compels us to reevaluate. But the fact that consciousness interprets the result does not automatically imply that it produces it.

(*I discuss this aspect more extensively in my other books, introducing the concept of the Field of Consciousness.*)

It must be stated openly: experimental physics has not yet established a single answer to the question of what "determines" the outcome.

According to the Bohrian approach, the demand to describe what happens before and during the measurement beyond certain limits may be misguided: the theory offers us rules for linking experimental setups and results— — not a visual account of what the object "does in itself" when no one is measuring it. Other perspectives, such as the one proposed by David Bohm in 1952, seek to preserve the idea of particles with a definite position, governed by a deeper quantum law. The many-worlds interpretation, put forward by Hugh Everett III in 1957, eliminates collapse and maintains that possible outcomes continue along different branches of reality. Theories of spontaneous collapse—formulated in an influential version by Gian Carlo Ghirardi, Alberto Rimini, and Tullio Weber in 1986—hypothesize, instead, that nature contains a real physical mechanism capable of making macroscopic situations definite.

We must not turn these hypotheses into a dizzying array of ideas. None of them justifies slogans such as "the mind creates everything" or "matter does not exist." If anything, they show us how difficult it is to reconcile three intuitions that are dear to us: that quantum theory is universally valid; that events have unique outcomes; and that the world does not depend on our observation to exist.

Perhaps the verb "to decide" is already misleading us. Deciding is a human act: it implies intention, choice, and alternatives that are understood and evaluated. An atom does not decide. A detector does not decide. Not even nature, in the sense in which the word is used in everyday language, seems to sit at a crossroads and choose its own destiny.

The more precise question, then, is another: what process transforms a description of possibilities into a concrete event? And we must add, with humility: are we certain that "transforms" is the right word, or are we trying to impose on reality a narrative that belongs solely to our mental habits?

The problem of measurement does not ask us to abandon reason. It asks us to make it more vigilant. It teaches us that between the world of equations and the world of experience there is not ne ly a rupture, but rather a transition that we have not yet learned to describe definitively.

Faced with that transition, the most fruitful stance is neither the arrogance of those who believe they have already solved the enigma, nor the mysticism of those who label everything they do not understand as "consciousness." It is a more demanding form of attention: recognizing that a simple click, in a silent laboratory, can contain an immense question about the relationship between possibility, matter, life, and presence.

Jakob von Uexküll's coin and the world that truly matters.

The cat imagined by Schrödinger is not merely a large object, placed in a somewhat grotesque manner at the end of a chain of microscopic events. It is not an atom magnified to the point of becoming domestic, complete with a tail and whiskers. It is a living being. And this detail, which is often treated as a mere narrative device, is perhaps the most unsettling aspect of the entire experiment.

A stone can fall, a pendulum can swing, a detector can register a signal. An organism, of course, also obeys the laws of physics: its body has mass, temperature, and chemical composition; its heart pumps blood according to principles that do not suspend the laws of fluid mechanics; its neurons transmit signals through electrical and molecular phenomena. There is no need to introduce, in order to explain life, a mysterious force that defies the laws of nature. Yet saying this is not enough.

To say that a cat is organized matter is true, but it is a truth that risks being too meager. It is a bit like saying that a Beethoven sonata is merely air set in motion by instruments made of wood and metal. It is not false. But it is not yet the sonata.

The problem posed by living beings does not, therefore, lie in the fact that they violate the laws of physics. Life does not violate the laws of nature: if it did, we would not be able to study it. The problem is different and more subtle. The laws of physics describe with extraordinary precision what happens to matter; but a living being is not merely matter to which things happen. It is matter that maintains a form, repairs its own damage, selects what is favorable to it, avoids what destroys it, develops, ages, suffers, and seeks.

A stone has no interest in remaining a stone. A cat, on the other hand, has a vital interest in staying alive.

This statement may seem obvious, but it contains a decisive difference. For a stone, there is no such thing as good or evil: it can break, wear away, or sink into the ground without anything—from its point of view—being lost. For an organism, on the other hand, there are favorable conditions and adverse conditions. Extreme cold, hunger, a wound, poison, or illness are not merely physical transformations: they are threats to the continuity of existence.

Living beings introduce into nature a dimension that a purely physical description, on its own, is not designed to express: the dimension of "for someone." The poison in Schrödinger's box has a chemical composition; it spreads according to certain laws; it acts on tissues and the nervous system through processes that can be studied. But for the cat, that poison is not merely a substance. It is what disrupts its world. Here we are no longer solely in the realm of physics. Not because physics fails, but because the question has changed.

Physics can tell us what happens to the molecules in a cell when it is exposed to a toxin. Biology can reconstruct the consequences on tissues and organs. Medicine can explain the sequence leading to the loss of vital functions. But no equation, in and of itself, captures the fact that for an animal, life is something to be preserved; that pain is not merely a nerve signal, but an experience of alarm; that death is not a different configuration of matter, but rather the end of the unity that held together a body, a sensibility, and a perspective on the world.

It is important not to overinterpret this observation. We are not demonstrating that life possesses a substance separate from matter, nor that consciousness must intervene to alter the laws of quantum physics. Quantum physics does not permit spiritual shortcuts. It does not prove that the soul causes possibilities to collapse, nor that every organism is a tiny divinity hidden within matter.

But it invites us, perhaps, to exercise caution: do not confuse the success of a description with the exhaustion of what is being described.

Schrödinger pondered a fundamental and unsettling question: how does an organism remain itself while being immersed in a world that tends toward disorder? A living body continuously exchanges matter with its environment. It breathes, feeds, eliminates waste, loses cells, and produces new ones. And yet it retains, for a certain time, a recognizable form and a surprising internal coherence.

In his writings, Schrödinger stated that an organism "feeds on negative entropy." The expression has remained famous, even though biologists today prefer to phrase it more precisely. It is not necessary to think that life contradicts the general tendency toward disorder: it does not. An organism maintains its organization solely because it absorbs energy and matter from the outside and returns heat and waste products to the environment. It lives by creating order locally, but it does so within an overall process that does not suspend any physical laws.

And yet Schrödinger's point retains its power. An organism is not an inert object that simply persists until some external cause alters it. It is a system that works ceaselessly to prevent its own dissolution. The its identity is not immobility: it is an unstable equilibrium, continually regained.

A cat sleeping on a windowsill appears still. But beneath that apparent stillness, a prodigious activity is taking place:

cells renewing themselves, proteins being built and broken down, nerve impulses, chemical exchanges, thermal regulation, defenses against external agents. Life does not coincide with a particular molecule, nor with a single organ, nor with a moment that can be captured in a photograph. It is a process. More than that: it is a form that sustains itself through change.

This is why living things force us to change our perspective. If we take a watch apart, we find its parts; if we reassemble those parts correctly, the watch starts working again. In an organism, the parts are essential, but they are not enough to explain the whole. A heart separated from the body is no longer a living heart in the full sense of the word; an isolated cell may survive for a time, but it has lost the network of relationships that assigned it a role within an organism.

This is not a concession to facile mystery. It is a fact that biology knows well: the same molecules can perform different functions depending on the tissue, the stage of development, the environment, and the organism's history. Living matter is not merely an assembly; it is matter embedded in relationships, regulations, and memories.

Jakob von Uexküll, an Estonian-German biologist active in the early decades of the 20th century, proposed a memorable image to help us understand this difference. While studying simple animals, including the tick, he observed that every living being inhabits its own meaningful environment, its own "Umwelt"—that is, a surrounding world selected according to what matters for its life.

The tick, which is nearly blind and nearly deaf, does not live in the same world as we do, even though it is in the same forest as us. For it, only a few signals are decisive: the smell of butyric acid emitted by mammals, the warmth of skin, and the texture of the surface on which it lands.

The rest may not exist at all—at least not as a relevant reality. Where we see a complex landscape—trees, colors, trails, wind, skies—the tick encounters a much more limited universe, made up of expectations and vital signals.

This does not mean that the tick invents the world in its own mind. It means, more simply, that an organism does not perceive reality as a neutral record. It selects, distinguishes, and reacts. What it encounters has value for it: nourishment or danger, shelter or threat, the possibility of mating or escape.

A cat is not a tick, of course. Its world is much richer: smells, movements, territories, habits, familiar presences, fear, play, hunger, sleep. But the point remains. A living being is not merely a point in space subjected to causes. It is a center of sensitivity and response. Even before tackling the even more difficult problem of human consciousness, life compels us to recognize that there is a difference between an event that happens and an event that happens to someone. It is here that Schrödinger's cat definitively ceases to be a stage prop.

When we say that the cat is alive or dead, we are not merely attributing a measurable property to a collection of atoms. We are recognizing a radical difference between two states. In the first case, there is still an organism capable of maintaining its unity, of feeling, of moving, of interacting with its environment. In the second, that unique organization has ceased to exist. The atoms remain, many structures remain, and for a certain time even chemical processes remain. But the living being no longer remains.

The difference is not imaginary because it cannot be reduced to a poetic phrase. Death can be established using rigorous medical criteria. And yet its meaning goes beyond the list of those criteria. For those who love an animal, for those who lose a loved one, for those who contemplate their own aging body, death is not merely a change of state.

It is the revelation that life was not something possessed, as one possesses an object. It was a fragile, temporary, unrepeatable occurrence.

Physics can take us a long way in describing the material conditions of this occurrence. It can tell us how molecules form, how energy flows, how systems stabilize or decay. But it was not designed to answer, on its own, the question that arises when a living being appears: what does it mean for this matter to have become a center of experience, vulnerability, and inquiry?

Perhaps the limitation lies not with physics, but with our expectation that a single language can say it all. We need physics so as not to lose ourselves in fantasies. We need biology to understand the concrete fabric of vital processes. But we also need words capable of speaking to experience, value, relationship, death, and care. Not because these words add a spiritual haze to what science has clarified, but because they name real aspects that change the meaning of the scene.

In Schrödinger's box, therefore, there is not only the technical problem of how a quantum possibility becomes an observable fact. There is an animal. And with that animal comes a question that no experimental apparatus can erase: what happens when matter does not merely exist, but begins— , in the elementary and silent way characteristic of life—to defend its own existence?

We do not yet have a definitive answer. Perhaps we will never have one in a form that can dispel all doubt. But we can at least avoid two opposing errors: turning life into a miracle that eludes all inquiry, or reducing it to a mechanism devoid of inner life, history, and meaning.

Between these two extremes lies a more challenging space. It is the space in which science continues to observe, measure, and explain; and we, without abandoning rigor, learn to recognize that a living being is not merely what it is made of. It is also the precarious and tenacious way in which what it is made of becomes a world.

IV. What is life?

"Living matter, while not defying the laws of physics established to date, probably implies 'other laws of physics' as yet unknown."
Erwin Schrödinger, What Is Life? (1944), Chapter 7

"Chromosomal structures contain, in a sort of written code, the entire blueprint for the individual's future development and its functioning in adulthood."
Erwin Schrödinger, What Is Life? (1944), Chapter 5

In February 1943, in Dublin, Erwin Schrödinger gave a series of public lectures. This time, he was not speaking about the behavior of electrons or the oddities of the atomic world. He was speaking about life. The question he had chosen as the title—"What Is Life?"—was as simple in form as it was mind-boggling in content.

What distinguishes a stone from a seed? A salt crystal from a cell? A complex molecule from an organism capable of growing, repairing itself, reproducing, and—at least in some forms—questioning its own existence? Schrödinger did not claim to offer a definitive definition. He was not an experimental biologist, and he knew it well. He was a theoretical physicist venturing respectfully into another's territory, yet posing a question that the biology of the time had not yet formulated with such clarity: How can living matter maintain its extraordinary organization in a universe that seems to tend, spontaneously, toward disorder?

The small book that emerged from those lectures, published in Cambridge in 1944, would exert an enormous influence. Not because it had solved the problem of life, nor even because it had anticipated molecular biology in every detail. Some of its hypotheses would prove to be incomplete or outdated. But Schrödinger saw with great clarity that life could not be understood merely as a set of chemical reactions. In living matter, there was a question of order, memory, the transmission of information, and continuity over time.

In this chapter, we will follow that thread. First, we will consider what physics has to say: life does not violate the laws of nature, but exists in a unique state, far from equilibrium. Then we will explore the intuition behind the "hereditary code," formulated before the structure of DNA was discovered. Finally, we will try to consider life not as an object placed in the world, but as an unceasing activity:

a process that sustains itself, renews itself, and, for a time, resists its own dissolution.

A living organism is not immobile. Even while it sleeps, even when it seems to be doing nothing, it is working against dispersion. Its cells exchange matter and energy with the environment; proteins are built and broken down; blood circulates; tissues repair themselves; the body's molecules enter and exit in a continuous transformation.

A stone can remain almost identical for centuries, but it does not have to do anything to achieve this. A living being, on the other hand, retains its form for some time only because it is constantly changing.

This is where one of the great ideas of nineteenth-century physics comes into play: entropy. Without getting bogged down in formulas, we can describe it as a measure of the dispersion of energy and the number of ways in which the components of a system can be distributed. A cup of hot coffee left on the table cools down; a drop of ink in water spreads out; an abandoned building tends to fall into disrepair. Not because there is some curse of disorder, but because more dispersed states are statistically much more likely than ordered states.

The second law of thermodynamics states, in general terms, that in an isolated system, entropy does not decrease. Energy tends to disperse, temperature differences level out, and structures capable of performing work are depleted. At the end of this ideal process, equilibrium would be reached: no flow, no significant differences, no spontaneous transformation.

But an organism is not an isolated system. It lives because it continuously exchanges with its surroundings. It breathes, eats, drinks, absorbs light, eliminates waste, and dissipates heat. It does not produce order out of nothing, nor does it defy the second law of thermodynamics. It does something more modest and more interesting: it locally preserves its own organization while increasing entropy in the environment.

Schrödinger expresses this with a formula destined to become famous: "*What an organism feeds on is negative entropy.*" (What Is Life? Chapter 6).

The formula for "negative entropy," or "negentropy," is evocative, but it must be handled with care. It does not mean that we consume a mysterious form of energy called negentropy. Schrödinger himself clarified that, in language closer to chemistry and biology, we could speak of free energy: energy available to fuel transformations, to build structures, and to maintain differences.

When we eat a piece of fruit, we do not merely introduce calories into our bodies. We introduce organized molecules, which in turn are produced by a long chain of processes: sunlight captured through photosynthesis, water absorbed by the roots, carbon dioxide transformed into sugars, and plant cells that have built and maintained their own structure. Our body uses that matter and that energy to continue being what it is, while releasing heat, carbon dioxide, and waste products into the environment.

Life, therefore, is not a magical interlude within physics. It is a particularly complex form of physics and chemistry that unfolds far from equilibrium. A flame, a whirlpool in water, or a hurricane are also temporary structures sustained by energy flows. And yet an organism does something more: it does not merely persist for a brief moment. It preserves a history, repairs itself, creates copies of itself, and transmits instructions.

Schrödinger had grasped the enigmatic nature of this temporary resistance to dissipation: "*It is by avoiding rapid decay into the inert state of 'equilibrium' that an organism appears so enigmatic.*" (What Is Life? Chapter 6)

The key word here is perhaps "rapid." Life, of course, also comes to an end. No individual organism remains outside of equilibrium forever. Death, viewed from the perspective of physics, is not an accident foreign to nature: it is the moment when that extraordinary ability to keep vital processes coordinated fails. Cells do not immediately cease to be matter; what ceases is their collective activity, the dynamic order that held them together. In this sense, living is not simply equivalent to possessing a body. It means maintaining a form through change.

A newborn, an adult, and an elderly person share very little identical matter: over the years, countless atoms and molecules have entered and left the body. Yet we say it is the same person. Not because the body is an immutable block, but because a continuity of organization, biological memory, relationships, and history runs through the change. Life forces us to correct one of our most deeply ingrained mental habits: the belief that identity depends on the permanence of matter.

Here, physics describes processes, exchanges, and energy constraints. When, on the other hand, we ask ourselves what it means—for us—to remain the same amid change, we enter a different realm: that of experience and philosophical reflection. It is important not to confuse these two planes. But it is equally important not to pretend that they are entirely unrelated. Schrödinger did not stop at thermodynamics. His second insight concerned heredity. How was it possible for such a complex organism to develop from a single initial cell? And how could the form of a living being—the arrangement of its traits, abilities, vulnerabilities, and —be passed down through generations with such remarkable fidelity?

In the 1940s, the structure of DNA was not yet known. Its genetic function was still a matter of debate. The experiments by Oswald Avery, Colin MacLeod, and

Maclyn McCarty, published in 1944, would provide strong evidence that DNA was the material responsible for hereditary transmission; but the decisive confirmation of its structure would not come until 1953, with the double-helix model proposed by James Watson and Francis Crick, which was also based on crucial data obtained by Rosalind Franklin and Maurice Wilkins.

Schrödinger could not have known all this. Yet he intuited that heredity must depend on a material structure capable of storing information in a stable, yet non-monotonous, way. An ordinary crystal, such as salt or quartz, regularly repeats the same atomic arrangement. It is stable but lacking in variety. To preserve the instructions necessary to build an organism, however, a material was needed that was both ordered and rich in differences.

Schrödinger called it an "aperiodic crystal": *"We believe that a gene—or perhaps the entire chromosomal fiber—is an aperiodic solid."* (What Is Life? Chapter 5).

This image should not be taken as a literal description of DNA as we know it today. DNA is a molecule with a regular structure—the double helix—but the sequence of its bases, denoted by the letters A, T, C, and G, is not a simple, uniform repetition. It is precisely the variation in the sequence that makes it possible to encode biological information.

Schrödinger's great insight was therefore this: hereditary material had to be not only chemically stable; it had to be capable of carrying a message. A message not in the human, in , intentional sense of the word—like a letter sent by one person to another—but in the sense of a configuration capable of producing ordered effects over time. In the pages of **What Is Life?**, Schrödinger uses a surprisingly modern expression: "code-script." In Chapter 5, he writes: *"Chromosomal structures contain, in a sort* *of code-script, the entire blueprint for the individual's future development and its functioning in adulthood.*"

Here, further clarification is needed. DNA is not a complete and rigid architectural blueprint, like the floor plan of a building. It does not contain, already drawn in miniature, a child's face, the path of a hand, a person's character, or the course of a disease. Genes operate within complex networks; they interact with other molecules, with the cellular environment, with nutrition, with chance, with experience, and, in more complex organisms, with the history of the entire body.

The genome is not a script that is simply recited. It is more like a set of possibilities and constraints, activated and modulated under ever-changing concrete conditions. Identical twins possess nearly the same genetic makeup, but they never live exactly the same biological life. Time, the environment, relationships, and even small differences in embryonic development introduce real variations. This does not diminish the importance of DNA; on the contrary, it makes it more interesting. Genetic material is not a predetermined fate, but a chemical memory that enters into dialogue with the world.

It was precisely this promise of a bridge between physics, chemistry, and biology that fascinated many young scientists in the postwar period. Francis Crick, who would later share the Nobel Prize with Watson for the discovery of the structure of DNA, recalled that Schrödinger's book was one of the works that steered him toward biology. Maurice Wilkins also recognized the importance of those pages. Not because the Austrian physicist had provided them with a solution, but because he had shown that the problem of heredity could be approached as a problem of structure and information.

Science often proceeds in this way: not only through exact answers, but thanks to questions asked at the right moment. Schrödinger, however, was far from proclaiming that life had now been explained. In one of the most cautious and profound passages of his book, he wrote: *"Living matter, while not evading the laws of physics established to date, probably implies 'other laws of physics' as yet unknown."*

This statement does not give us license to ascribe to life whatever mystery we desire: a measurable soul, a hidden energy, a force that contradicts chemistry, or a cosmic consciousness demonstrated in the laboratory. Schrödinger is not saying this. He is saying something more rigorous: the known laws may not be sufficient, at a certain point in history, to understand all levels of matter's organization.

Contemporary biology has made strides that Schrödinger could not have imagined: it has sequenced genomes, observed proteins at work, reconstructed cellular regulatory networks, studied the role of the microbes that inhabit our bodies, and identified mechanisms by which the environment can influence gene expression. Yet the question "What is life?" remains unresolved.

We can list recurring characteristics: metabolism, the ability to maintain a boundary with the outside world, reproduction, evolution, and response to the environment. But every definition encounters cases that challenge it. Viruses, for example, possess genetic material and evolve, but they lack an autonomous metabolism: to replicate, they must enter the cells of other organisms. Are they alive? It depends on the criterion used. And the question, rather than fading away, becomes more precise.

Perhaps the point is not to find a single property, a secret substance called "life," hidden somewhere within cells. Perhaps life is recognizable above all as an organized relationship among many activities: a membrane that separates and connects; a metabolism that transforms; an inheritance that preserves and varies; a capacity to respond; a balance that is never static, but rather continuous regulation.

An organism is a boundary, but not a fortress. If it stopped exchanging with the world, it would die. If it ceased to renew itself, it would degrade. If it had no memory, it could not maintain its form. To live means, in a certain sense, to be open without dissolving.

This is perhaps the most fruitful legacy of **What Is Life?**. Schrödinger does not leave us a definitive formula; he teaches us to view the living as a form of fragile, active, and temporary order. Not an object we possess, not a machine reducible to a single part, not a magic that suspends nature. Rather, a process in which matter, energy, and information are held together for a limited time.

And that limited time is not a minor detail. It is the very condition that makes life seem precious to us. Every organism lives because it manages, for a time, to escape dispersion; every organism dies because that escape cannot last indefinitely. Between these two poles—continuity and dissolution—the story of every living being takes shape.

Physics can describe an essential part of this story. It can show us why life requires flows, energy, and exchanges. Biology can reveal, with increasing precision, the mechanisms of transmission and transformation. But one question remains open— —and it is not confined to laboratories: What does it mean to be that portion of matter which, for a brief time, not only organizes itself but also feels, remembers, and asks questions?

It is not yet time to answer. It is time to recognize that life, before being a concept, is the very starting point of all our questions.

The concept of "negentropy."

Let's look at an organism without the habit that renders the most surprising things invisible. A tree grows, heals a wound in its bark, turns its leaves toward the light, and transports water from its roots to its crown. A human body heals itself, replaces cells, regulates its temperature, and transforms food into movement, memory, and heat. Even during sleep, when consciousness withdraws from the scene, the body never stops working: it breathes, circulates, selects, eliminates, and rebuilds.

This stability is not immobility. On the contrary: a living being preserves its form only because it is undergoing ceaseless transformation. A stone can remain in the same place for centuries without having to do anything; an organism, on the other hand, remains itself only by constantly changing.

Here a term from physics comes into play that is often invoked too readily: entropy. In everyday language, we call it disorder. The image is useful, provided it is not taken literally. In physics, entropy primarily concerns the tendency of energy to disperse, of differences to diminish, and of systems to reach more uniform and more probable states. A cup of tea left on a table cools down; conversely, we never see the heat in the room spontaneously gather in the cup to the point of bringing it to a boil. Perfume sprayed in a corner spreads through the air; it does not return to the bottle on its own. The wood in an abandoned house rots, iron rusts, and cracks widen. There is no malevolent force of chaos at work: there is simply the fact that the most ly dispersed states are immensely easier to achieve than concentrated and ordered states.

The second law of thermodynamics, formulated in the nineteenth century by physicists such as Rudolf Clausius and William Thomson, Lord Kelvin, states in general terms that the entropy of an isolated system does not decrease. Left to itself, a system tends toward equilibrium: temperature differences diminish, pressures equalize, and the energy capable of producing change becomes increasingly scarce.

Equilibrium, from a physical standpoint, is not a catastrophe. It is a state of stillness. But it is a stillness in which nothing spontaneous happens anymore. Life, on the other hand, happens precisely because it is not in equilibrium.

An organism is not a closed system, separated from the rest of the world like a sealed box. It breathes, eats, drinks, absorbs light, expels substances, and dissipates heat. Its apparent autonomy depends on an uninterrupted relationship with its surroundings. It does not produce order out of nothing, nor does it violate the second law of thermodynamics. It does something more concrete: it maintains a local order by releasing a greater amount of entropy into the environment.

This is an essential point, because it protects us from one of the most widespread distortions of scientific thought. Life is not a miraculous exception to the laws of matter. It is not a small realm exempt from physics. It is a physical-chemical process of extraordinary complexity that manages, for a certain time, to delay its own slide toward equilibrium thanks to a continuous exchange with the world. Schrödinger condensed this idea into a formula *that* has become famous: *"What an organism feeds on is negative entropy.*" The expression is striking because it seems almost contradictory. How can "negative" entropy exist? And above all: what does it mean for an organism to feed o ? The word *negentropy*—a shortened form of *"negative* entropy" that became common in subsequent years—does not refer to a mysterious substance, a sort of invisible fuel for life. It is not something that can be poured into a glass or measured like a vitamin. Schrödinger himself, a few lines later, suggested a less paradoxical formulation: living organisms feed on free energy—that is, energy available to produce transformations.

A fully charged battery possesses free energy; a discharged battery possesses much less. Water collected upstream of a dam possesses a difference in elevation relative to the valley, and it is precisely this difference that can be harnessed to produce work. A warm room next to a cold one possesses a temperature difference; when the two temperatures equalize, that potential is exhausted. Life requires differences: differences in concentration, temperature, chemical composition, and energy. If everything were perfectly uniform, there would be no flows; and without flows, there would be no metabolism.

Consider a leaf exposed to the sun. The light arriving from the Sun is concentrated energy, originating from a very hot source. The plant uses a small portion of that energy to build complex molecules from water and carbon dioxide. It does not "create" energy: it transforms it. In doing so, it produces and dissipates heat; the planet returns energy to space in the form of infrared radiation, in greater quantity than it received. On balance, entropy increases. But locally—in the leaf, the trunk, the seed—an ordered, temporary structure takes shape.

When we eat a piece of fruit, therefore, we do not merely ingest calories. We receive the condensed result of a cosmic and terrestrial history: sunlight captured through photosynthesis, water drawn up from the roots, mineral salts and absorbed from the soil, molecules built and organized within plant cells. Our body uses a portion of this matter and energy to maintain its own structures, while releasing heat, carbon dioxide, water, and waste products back into the environment.

The order of life always comes at a cost. Every gesture, every thought sustained by a living brain, every heartbeat, every healing of a wound involves transformations that increase entropy outside the organism. Life does not eliminate dispersion: it shifts it, manages it, and accelerates it elsewhere in order to preserve its own form for a little while.

This dynamic can also be observed in non-living phenomena. In the early 20th century, the French physicist Henri Bénard studied a thin layer of liquid heated from below. When the temperature difference exceeds a certain threshold, the liquid no longer simply transfers heat randomly: regular hexagonal convection cells appear, now known as Bénard cells. The fluid rises in some places and sinks in others; an orderly pattern emerges in the movement.

It is a fascinating phenomenon: order can emerge even as energy flows and dissipates. But we must not take the analogy too far. Bénard cells are not alive; they do not repair themselves, they do not feed, and they do not retain a memory of their own history. Yet they reveal something important: disorder is not always the opposite of all form. Under certain conditions, far from equilibrium, the flows themselves can generate ordered configurations.

An organism is incomparably more complex than a vortex or a convection cell. A flame possesses a recognizable shape as long as it receives fuel and oxygen; a hurricane retains an impressive structure for a certain time thanks to thermal and l differences in the atmosphere. But neither the flame nor the hurricane constructs its own internal structure made up of membranes, regulated exchanges, repairs, and coordinated rhythms. The living is not merely a form maintained by a flow: it is a form that uses the flow to sustain itself.

This is why Schrödinger could write: *"It is by avoiding rapid decay into the inert state of 'equilibrium' that an organism appears so enigmatic.*"

The key word here is perhaps "rapid." Even an organism, of course, cannot escape equilibrium forever. Negentropy is not a promise of immortality. Every living body is temporary; every equilibrium achieved must be regained. When the network of exchanges, repairs, and regulations can no longer sustain itself, the body's matter remains—at least for a time—but the organization that held it together as a living whole ceases to exist.

From this perspective, death is not the arrival of disorder in a reality that was previously free of it. It is the end of an active, fragile, and continuous resistance against dispersion. Perhaps this is where the notion of negentropy ceases to be merely a technical concept and becomes a lens capable of transforming our perspective. To live does not mean to possess an immobile form. It means to inhabit an unstable equilibrium, to preserve it for a time, and to renew it ceaselessly through exchanges with that which we are not.

Life is not opposed to nature. It is one of nature's most subtle forms: matter that, for a limited time, manages to gather dispersed energy, to shape it into structure, to delay its own dissolution, and to transform this temporary victory over uniformity into breath, growth, sensitivity, and history.

The Code of Life.

When Erwin Schrödinger published **What Is Life?** in 1944, no one had yet seen the famous double helix of DNA. The structure of the molecule that we now almost instinctively associate with genes, family inheritance, diseases, and biotechnology would not be proposed until nine years later, in 1953, by James Watson and Francis Crick in Cambridge.

Yet Schrödinger had already identified the crux of the problem: if an organism manages to preserve and transmit its form across generations, there must exist, within living matter, a mechanism capable of storing an enormous amount of instructions in an incredibly small space.

He did not use the contemporary language of molecular genetics, because that discipline, in the full sense of the term, had yet to emerge. But he posed a decisive question: what physical structure allows a living being to remain true to itself without being identical to any other?

An acorn does not contain a miniature oak tree, as if the tree were already folded within it on a smaller scale. And yet it carries within it something that, upon encountering the soil, water, light, the seasons, and time, will make it possible for roots, a trunk, leaves, and acorns to grow. In the same way, an entire organism—composed of billions of different cells: skin cells, neurons, muscles, and blood cells—can develop from a single fertilized cell. How can matter that is initially almost indistinct— —give rise to such organized diversity? This was the concrete wonder that Schrödinger sought to unravel.

Biologists in the early 20th century already knew that heredity depended on discrete units—genes. Gregor Mendel, the Augustinian monk who had studied pea plants in the garden of St. Thomas Abbey in Brno, in what is now the Czech Republic, had published the results of his experiments in 1866. His work remained virtually ignored for decades before being rediscovered in 1900. From then on, the idea that certain traits were transmitted according to recognizable rules gradually gained acceptance. But one question remained unanswered: what are genes made of?

For a long time, many biologists had thought that the most plausible answer was proteins. They seemed like molecules varied and sophisticated enough to hold the secret of life. DNA, on the other hand, appeared to many to be a relatively simple, almost monotonous substance: a long chain composed of a few recurring elements. It did not seem to possess the richness necessary to explain the astonishing complexity of an organism.

In 1944, the very same year **What Is Life?** was published, a group of researchers at the *Rockefeller Institute for Medical Research* in New York published a finding that would slowly change the picture. Oswald Avery, Colin MacLeod, and Maclyn McCarty were studying a bacterium responsible for certain forms of pneumonia, **Streptococcus pneumoniae**. They demonstrated that a hereditary trait could be passed from one bacterial strain to another via DNA.

The experiment was not immediately accepted as a definitive revelation. This caution was understandable: in research, an important finding must be confirmed, discussed, and subjected to further testing. But the direction had now been set. DNA was not merely a “ ” supporting material within the cell nucleus; it could be the vehicle of hereditary information.

Schrödinger did not base his reasoning on the experiments by Avery and his collaborators: his lectures in Dublin had been delivered in February 1943, before the American article was published. But his book and those experiments belong to the same historical moment—that threshold at which life was beginning to be examined also as a problem of information.

Here it is worth pausing for a moment, because the word "information" can be misleading.

In everyday language, information is a piece of news, a message, or a sentence that someone directs at someone else. In a cell, there is no author who writes and no reader who understands in the human sense of the term. DNA is a molecule: a long sequence of chemical components. Yet the order of those components matters. It matters enormously.

We can imagine an alphabet consisting of four letters. In DNA, these "letters" are four nitrogenous bases: adenine, thymine, cytosine, and guanine, denoted by the initials A, T, C, and G. What makes the difference is not merely the presence of these elements, but their sequence. As in human languages, a few letters can form words, and a few words can compose radically different texts. On an incomparably larger and more precise scale, variations in the DNA sequence can alter the instructions by which cells produce proteins, regulate processes, and respond to the environment.

Schrödinger called the hereditary structure a *"code-script.*" It is a unique and apt expression. It speaks not only of a code—that is, a rule of correspondence—but also of a script. It evokes something that can be preserved, copied, transmitted, and, at times, altered.

We must not attribute to Schrödinger what he could not have known. He did not foresee the double helix in its concrete form; he did not identify the four bases of DNA as the genetic alphabet; he did not explain the mechanism by which genetic sequences direct protein synthesis. But he intuited with extraordinary clarity that heredity must depend on a structure stable enough to endure over time and rich enough to contain an enormous variety of possible arrangements.

To describe it, he introduced an image that would become famous: the "aperiodic crystal."

A common crystal, such as salt or quartz, has an ordered structure that repeats regularly. If we could observe its internal architecture, we would see a repeating geometric pattern. This regularity gives the crystal stability, but it prevents it from containing an unlimited variety of messages: a pattern that repeats exactly the same way conveys very little.

An organism, on the other hand, must reconcile two seemingly contradictory needs. It needs stability, because hereditary information cannot be lost with each generation. But it also needs variety, because the differences between species, individual diversity, and the very possibility of evolution depend on it.

The aperiodic crystal envisioned by Schrödinger would have possessed precisely these characteristics: a solid material structure, yet with a non-repetitive sequence capable of preserving a complex configuration. Not a random mosaic, but an order that does not merely repeat a single pattern.

When Watson and Crick published their short article titled *"A Structure for Deoxyribose Nucleic* Acid" in *Nature* on April 25, 1953, the double helix gave this insight a concrete physical form. Two complementary strands wind around each other; the bases— —pair according to precise rules—adenine with thymine, cytosine with guanine—and it is precisely this complementarity that makes the replication of genetic material possible.

The famous closing sentence of Watson and Crick's article was terse, almost restrained: *"It has not escaped our notice that the specific pairing we have postulated immediately suggests a possible mechanism for the replication of genetic material.*" It was a short sentence, but it ushered in a new era.

Behind that result were not just two young researchers at Cambridge. There was the experimental work of Rosalind Franklin and Maurice Wilkins at King's College London, particularly their X-ray diffraction studies of DNA fibers. The photograph known as "Photo 51," taken in May 1952 by graduate student Raymond Gosling under the supervision of Rosalind Franklin, provided a crucial visual clue to the molecule's helical shape. The story of that discovery is also a story of difficult relationships, belated recognition, and a scientific community that has not always been able to distribute credit fairly.

Rosalind Franklin died in 1958, at the age of thirty-seven, likely from ovarian cancer. Four years later, in 1962, Watson, Crick, and Wilkins received the Nobel Prize in Physiology or Medicine. The Nobel Prize is not awarded posthumously, but this formal detail does not fully address the issue: Franklin's story continues to remind us that scientific knowledge never arises from an isolated mind, but rather from a web of techniques, intuitions, rivalries, acts of collaboration, and sometimes injustices.

Francis Crick would later openly acknowledge just how important *What Is Life?* had been for his generation. In his 1988 autobiography, *What Mad Pursuit*, he recalled that Schrödinger's essay exerted a strong influence on young physicists eager to enter the field of biology. For many of them, life no longer appeared as a purely descriptive subject, left to naturalists; it became a problem demanding a new alliance between physics, chemistry, and biology.

The idea of the "code of life," however, has also met with an ambiguous fate. It has been a powerful metaphor, capable of guiding decisive discoveries. But, if taken too literally, it risks creating a misunderstanding: that of imagining the genome as a person's complete and immutable blueprint.

We are not simply the result of a prewritten script.

DNA makes certain structures possible, regulates the production of molecules, and conveys constraints and potential. But a gene does not operate in a vacuum. It functions within a cell; the cell lives within tissue; the tissue belongs to an organism; the organism grows in an environment, encountering nourishment, diseases, relationships, languages, habits, and events. The same genetic sequence can be activated or silenced in different ways in different cells and at different stages of life.

A liver cell and a neuron contain, generally speaking, the same genetic makeup. Yet they do not become the same thing. Their difference stems from the fact that they read—to use yet another inevitable metaphor—different parts of that same chemical library. Some chapters are consulted, others remain closed; some instructions come into play during embryonic development, others during an immune response, and still others during aging.

Contemporary biology has made this picture more complex thanks to the study of epigenetics: the set of mechanisms that can modify gene activity without changing the DNA sequence. This does not mean—as is sometimes claimed with excessive casualness—that every personal experience rewrites genes or that the mind can arbitrarily determine its own biology. It does mean, however, that there are many levels of regulation and response between the genetic sequence and the actual life of an organism.

The code is not a destiny in the theatrical sense of the term. It is more like a grammar: it makes certain sentences possible, prevents others, but does not decide on its own what speech will be uttered.

Here, biology has the final say. And that is enough to correct two opposing oversimplifications. The first is genetic determinism: the idea that everything we are—character, abilities, suffering, choices—is definitively contained within the DNA sequence. The second is the opposite illusion, according to which biology does not matter, and willpower alone is enough to free us from every inherited condition.

Neither of these views holds up. We come into the world with a body that has a history older than our own, a memory accumulated over millions of years of evolution. But this memory does not speak with a single voice; it enters into a relationship with the world, and the world, in turn, enters into us.

At this point, we can cautiously venture toward a broader meaning. Not yet a philosophical conclusion, and certainly not a spiritual demonstration. Merely an image to inhabit: at the heart of every cell lies matter capable of preserving a trace, transmitting it, and altering it. A memory without consciousness, a text without an individual author, which nevertheless makes possible the emergence of beings capable of remembering, narrating, and interpreting.

Schrödinger had not discovered DNA. He had, however, seen that life demands something surprising from matter: not only energy and structure, but continuity of form across time. His "aperiodic crystal" was an attempt to name this unstable fidelity: matter ordered enough to remember, open enough to change.

Perhaps the mystery is not diminished by knowing that part of our heritage is preserved in a molecular sequence. Rather, the nature of the question changes. We no longer ask only what matter we are made of. We ask how matter can become memory; how a chemical memory can participate in the birth of a body; and how, within that body, someone might ultimately emerge who asks what the meaning of their own origin is.

To live is to continually remake oneself.

At first glance, an organism seems like a thing. It has a shape, a weight, a skin that separates it from the outside world; it can be photographed, measured, cataloged. A tree stands there in the park; a dog crosses the street; a body rests in a chair. Language itself invites us to think of the living as an object among objects.

But this impression, however natural, is incomplete. An organism does not truly resemble a stone, a coin, or a statue. It is not a stable substance that, every now and then, undergoes some change. Rather, it is an activity that manages, for a certain time, to give itself form. It is not something that simply is: it is something that is continually happening.

Our body demonstrates this with an almost disconcerting subtlety. We breathe, assimilate, excrete, repair, remember, and react. The molecules that make us up come and go ceaselessly; skin cells renew themselves, blood is produced in the bone marrow, and the intestine rapidly rebuilds its lining. Not everything in the body is replaced at the same rate—some nerve cells, for example, can remain with us for many years—but nothing in a living being simply remains static. What we call "I," from a bodily perspective, is a continuity maintained through ceaseless renewal.

We could say that an organism does not preserve its identity in spite of change, but thanks to change. If it were to cease transforming matter and energy, distinguishing what it needs from what harms it, repairing its own structures, and responding to the environment, it would not become a quieter version of itself: it would cease to live.

It is here that the question posed by Schrödinger in *What Is Life?* takes on a depth that goes beyond the nascent molecular biology of his time. He wondered how an organism could escape, locally and temporarily, the tendency toward disorder. But the answer cannot be imagined as the defense of a small, immobile fortress against external chaos. Life is not a rigid order. It is an order in motion, an unstable equilibrium that is continually recreated.

A heart, for example, is not an object that possesses a beat as a secondary characteristic. Its beating is part of what it is. A heart that does not beat retains a recognizable anatomical form for some time, but no longer performs the function that makes it part of a living being. Similarly, a lung is not merely a collection of tissues: it is the possibility of breathing; an immune system is not a collection of cells: it is recognition, tolerance, defense, and memory. In living beings, structure and action cannot be separated too easily.

This observation may seem obvious, but it carries an important implication. To understand a cell, an organism, or a mind, it is not enough to simply list the parts that compose them. Knowing all the molecules in a cell does not yet amount to understanding how those molecules, together, manage to maintain a membrane, extract energy, correct errors, respond to a signal, divide, or refrain from dividing. The parts are indispensable; without them, there would be no life. But the parts, on their own, do not reveal how they interact with one another.

This is not to deny the value of physical and chemical explanations. On the contrary: contemporary biology thrives on the precision with which we can describe reactions, proteins, genes, membranes, and electrical signals. To say that life is more than the sum of its components does not mean invoking a mysterious force acting from outside matter. It means recognizing that the same matter can take on radically different configurations depending on the organization that permeates it.

A pile of bricks is not a house. But neither is a house merely a well-arranged pile of bricks: it is a set of relationships, functions, livable spaces, distributed weights, and openings to the outside. In an organism, however, the matter is even more radical: the "house" continually builds and repairs parts of itself, modifies its interactions with the environment, and in doing so changes without ceasing to be recognizably itself.

The Chilean biologist Humberto Maturana and his best-known collaborator, Francisco Varela, sought to articulate this insight in the early 1970s. In 1972, in Santiago, Chile, they proposed the term *autopoiesis*, from the Greek: *"self-production*." By this word, they did not mean to suggest that an organism creates itself out of nothing. Every living being depends on a world that precedes it: on air, water, food, light, other organisms, and the evolutionary history that made it possible. The idea was more precise: a living system is a network of processes that continuously produces and regenerates the components necessary for its own existence as a distinct unit.

From this perspective, a cell is not a bag filled with substances. Its membrane is not merely a passive boundary, like the glass of a bottle. It is an active surface: it allows some substances to pass through, expels others, receives signals, and maintains differences in concentration and without which the cell could not function. And the membrane itself is produced and repaired by the activities taking place within. The cell maintains its boundary, but that boundary is not a wall: it is a threshold of exchange.

This image can also help us correct a widespread misunderstanding. When we say that an organism is "separated" from its environment, we are not saying that it can exist without an environment. On the contrary: an organism is distinct precisely because it engages in selective exchanges with its surroundings. It is a living boundary, not isolation.

Consider a tree. Its form seems motionless compared to our haste, yet no living being depends so visibly on what passes beyond it. The roots harbor fungi and bacteria; the leaves transform solar energy; water rises through the tissues; carbon from the atmosphere enters the trunk's tissue. A tree does not merely stand in the landscape: the landscape passes through it. And yet this does not mean the tree dissolves into the environment. It retains its own form, history, and rhythm.

The French naturalist and philosopher Georges Canguilhem, who taught for many years at the Sorbonne and was Michel Foucault's mentor, insisted on an essential point: life is not indifferent to what happens to it. An organism does not passively receive the impacts of the world. For it, some conditions are favorable, others harmful; some openings are possibilities, others threats. Canguilhem summed up this idea in a brief phrase: *"Life is polarity.*" To live means to distinguish, evaluate, and prefer in an elementary and biological sense of the term: to seek what allows one to continue, to avoid what leads to destruction.

We must not attribute human consciousness or reflective intention to a cell. That would be a poetic shortcut, not an explanation. But we can say that even the simplest form of life exhibits a direction: it tends to preserve itself, to protect itself, to reproduce, and to modify its behavior in response to external conditions. Life is not merely organized matter;

it is matter that, through its own organization, actively maintains the conditions for its own continuity.

This helps us understand why illness cannot always be reduced to the presence of a single "broken part." Of course, there are molecular defects, mutations, infections, and localized lesions. But health concerns the entire interplay of regulatory mechanisms through which an organism copes with variations, injuries, fever, fatigue, and aging. The physician does not treat a gene in the abstract nor an organ isolated from the body: he or she treats, as far as possible, a living being situated within a specific context, with a biological and personal history.

Even cancer, which rightly strikes us as one of the most terrible diseases, reveals the complexity of this interplay. In many cases, it is not the arrival of foreign matter, but the loss of coordination among the organism's own cells: cells that continue to proliferate without responding to the signals that normally integrate them into the tissue. It is a painful but instructive example. Life is not simply the growth of its parts. A cell can be biologically active and, precisely for this reason, contribute to the ruin of the whole of which it was a part. For a complex organism, living also means maintaining a balance among many different activities.

At this point, the limitation of an overly mechanical representation becomes clearer. A machine can be taken apart and put back together: its parts, at least in principle, retain their identity outside the whole. An organism, on the other hand, exists only within a web of interconnected processes. If we separate an organ from the body—and unless we provide it with very specific artificial conditions—it rapidly loses its ability to function. If we separate a cell from its environment, we discover that the environment was not merely a backdrop: it was integral to the cell's very existence.

The question becomes even more fascinating when we consider viruses. In 1935, the American biochemist Wendell Meredith Stanley succeeded in crystallizing the tobacco mosaic virus, demonstrating that it could take the form of an ordered chemical structure. Eleven years later, this discovery earned him the Nobel Prize in Chemistry. But that very success made one question even more difficult to answer: Is a virus alive?

Outside a host cell, a virus has no metabolism of its own; it does not produce energy, does not repair itself, and does not reproduce autonomously. Yet, once it enters a suitable cell, it can trigger a sequence of events capable of generating new viral particles. The virus seems to straddle the boundary between object and process. As a particle, it can be isolated and analyzed; as a biological phenomenon, it appears only in relation to a living cell. There is no universally accepted answer to the question of whether it is "alive" or not. But this very uncertainty is instructive: it reminds us that life cannot always be confined by a clear-cut definition, like some sort of stamp affixed to things.

Synthetic biology, too—often discussed in overly dramatic terms—confirms the same lesson. In 2010 in Rockville, Maryland, a group of researchers led by Craig Venter announced a cell controlled by a genome synthesized in the laboratory. Life had not been created out of nothing, as some newspaper headlines suggested. The artificial genome, designed based on that of the bacterium "*Mycoplasma mycoides*," was inserted into a recipient cell of "*Mycoplasma capricolum*," which already possessed membranes, ribosomes, enzymes, and all the material apparatus necessary to read that genetic material.

The experiment was extraordinary, but not because it had solved the mystery of life. It was extraordinary because it demonstrated how much genetic information needs a living context to become active. A DNA sequence, on its own, does not breathe, does not grow, does not sense a chemical gradient, and does not rebuild a membrane. The biological significance of the code arises from the coordinated work of an entire cell. The gene matters immensely, but it is not a sovereign that rules alone.

Here, Schrödinger's lesson retains its power, even though subsequent biology has corrected, expanded, and made his intuition more concrete. He had seen that in living organisms, matter and order cannot be conceived of separately. Today we can add that neither matter, information, nor the environment can be separated without losing something crucial. Life takes shape in the encounter between these three elements: a material structure, a memory capable of being transmitted and transformed, and a world with which a continuous exchange takes place.

On the level of physics, all of this remains within nature: atoms, molecules, energy, reactions, fields, and flows. There is no need to introduce a vital principle foreign to the laws of physics. But on the level of understanding, living things compel us to change our perspective. It is not enough to ask: What is it made of? We must ask: What process holds these parts together? What relationships make their cooperation possible? What happens when that cooperation is interrupted or transformed?

And on a more human level, almost inevitably, this image of life concerns us. We are accustomed to seeking with , an unchanging core that defines us: an essence, a character fixed once and for all, an identity preserved like a precious object. Biology does not allow us to conclude that we are "merely" flow, nor that the self is an illusion without substance. It does, however, suggest caution: perhaps continuity does not coincide with immobility.

A living being remains itself not because it retains everything, but because it undergoes transformations without completely losing its thread. In the body, this continuity consists of renewal and regulation; in personal life, of memory, bonds, wounds, choices, and language. We are not stones removed from time. We are vulnerable forms that time, for a time, manages to hold together.

Perhaps this is one of the most understated and most fruitful images that the science of life can offer us: to live does not mean to possess a definitive form. It means participating in a ceaseless process of composition, repair, and transformation. Not something closed in on itself, but a tapestry that continues to take shape.

V. The Self, the Cell, and the Sea of Being.

"The total number of minds is one."
Erwin Schrödinger, *Mind and Matter* (1958).

"The order and connection of ideas is the same as the order and connection of things."
Baruch Spinoza, Ethics, Part II, Proposition 7 (published posthumously in 1677).

"The many become one, and are increased by one."
Alfred North Whitehead, Process and Reality (1929).

"A thing is right when it tends to preserve the integrity, stability, and beauty of the biotic community. It is wrong when it tends toward the opposite."
Aldo Leopold, A Sand County Almanac (1949).

"We are not individuals: we are colonies."
Lewis Thomas, *The Lives of a Cell: Notes of a Biology Watcher* (1974).

After examining life as a process capable of maintaining and renewing its own order, an even more radical question arises: where does a living being truly end? Does it end at the skin, at the membrane of a cell, at the perimeter of the body, at the boundaries of the consciousness we call "I"? Or are these boundaries—though real and necessary in everyday life—merely provisional lines drawn within a vaster reality?

Schrödinger did not offer a scientific proof of the unity of all things. It would be incorrect to attribute this to him. Physics can describe relationships, processes, and transformations; it cannot automatically convert a spiritual vision into a laboratory result. Yet, alongside the rigorous physicist, there lived within him a thinker who was convinced that the absolute separation between mind and matter, subject and world, individual and totality, was too crude a simplification to account for our experience. In this chapter, we will explore that question without turning it into dogma: what changes if life is not merely composed of parts, but is the way in which the parts relate to one another?

The Unity of the Living.

There are words we use so naturally that we forget how strange they are. One of these is "individual." The word suggests something that cannot be divided: a compact, separate, self-sufficient unit. We say "I" and almost spontaneously imagine a stable center: a body bounded by skin, a personal history, a memory, a consciousness observing the world from within.

Yet, one need only look a little more closely to realize that the living individual is never isolated in the full sense of the term. It breathes air it did not produce; it feeds on matter that moments ago belonged to a plant, an animal, a fungus, or another organism; it depends on temperature, gravity, and its environment; it lives amid bacteria, viruses, cells, languages, learned gestures, and relationships. The body we call "mine" is a necessary boundary, but it is not a wall.

Even on a biological level, our apparent compactness proves to be more complex than it seems. The human body hosts immense communities of microorganisms, especially in the gut, on the skin, and in the respiratory tract. They are not merely accidental guests: they participate in digestion, interact with the immune system, and influence essential processes of our existence. The American biologist and physician Lewis Thomas, author of the 1974 *book *The Lives of a* Cell*, urged us to change our perspective. When looking at the human organism, he did not see a machine perfectly closed in on itself, but an * *—a lived, interwoven reality composed of ancient collaborations.

Of course, we must not fall into the opposite extreme: we are not "merely" microbial colonies, just as we are not merely a collection of molecules. We can speak, remember, promise, suffer, and choose; we possess a personal continuity and a sense of responsibility that do not disappear simply because our bodies are made up of relationships. But this complexity compels us to reframe the question. Is the self truly a fortress, or is it a center of experience that emerges from a larger fabric?

For Erwin Schrödinger, this question was not a philosophical embellishment tacked onto physics in moments of leisure. It was a central problem. He had been one of the founders of wave mechanics, the theory that in 1926 gave a new mathematical form to the description of the atomic world. But just as his physics was helping to transform the way we think about matter, Schrödinger was developing a more general concern: the modern scientific framework, so powerful in explaining phenomena, risked presenting the world as an expanse of separate objects, observed by a subject who remains outside the scene.

This image had deep roots in European history. In the seventeenth century, René Descartes had drawn a very clear distinction between the *"res cogitans,"* the thinking reality, and the *"res extensa,"* the material reality extending through space. It was a fruitful distinction for the birth of modern science: it allowed for the study of bodies, celestial bodies, and natural phenomena as objects subject to laws, without having to resolve questions of faith, the soul, or destiny each time. But, as the centuries passed, what had begun as a methodological distinction risked becoming an absolute division: on one side, matter; on the other, the mind; on one side, the body; on the other, experience; on one side, the external world; on the other, the self that observes it.

Schrödinger did not deny the difference between a thought and a stone, between pain and an electrical impulse, between inner experience and the physical description of a brain. It would be a caricature to attribute such naivety to him. His point was more subtle: perhaps the difference does not imply a definitive rupture. Perhaps mind and matter are not two alien realms miraculously forced to meet, but two ways in which the same reality manifests itself and is known.

Here it is essential to distinguish between levels. Quantum physics, as physics, does not prove that “everything is one” in the spiritual or religious sense of the phrase. The equations do not speak of the One of the wisdom traditions; they describe with extraordinary precision states, probabilities, interactions, and measurable quantities. They do not justify shortcuts such as: “since the quantum world is strange, then every metaphysical belief is scientifically proven.” This would be a confusion, not a synthesis.

But physics can challenge certain overly simplistic notions. It can show us, for example, that matter is not a collection of small, rigid, and independent objects—like balls colliding in a vacuum—that the description of a system depends on the relationships that constitute it; and that the observer is not always easily separable, in terms of measurement, from the apparatus with which they observe. No binding philosophy follows from this. What does follow, however, is a sense of caution: perhaps our habitual way of dividing reality into distinct compartments is not the final word.

Schrödinger felt this caution with particular intensity when he spoke of consciousness. In *_Mind and_ Matter*, a collection of lectures published in 1958, he wrote a sentence as brief as it is provocative: _“The total number of minds is one._” It is a l statement that may irritate, fascinate, or be misunderstood. It does not mean that people are interchangeable, nor that one person's pain should be treated as irrelevant because, after all, "we are all the same thing." Nor does it mean that Schrödinger solved the problem of consciousness. Rather, it means that he rejected the idea of a multiplicity of completely separate consciousnesses, each closed in on itself like windowless monads.

His insight was close to an age-old question: when we say "I," what exactly are we talking about? An immutable entity? A separate individual soul? An unbroken history? Or a process of memory, perception, and relationship that, while possessing its own undeniable uniqueness, lacks the kind of absolute isolation we tend to attribute to it?

Modern biology, in language very different from Schrödinger's, offers some examples that make this question less abstract. Consider mitochondria, tiny structures found in the cells of animals, plants, and fungi. They produce much of the energy that allows our cells to live. The theory of endosymbiosis, proposed and defended with great tenacity by the American biologist Lynn Margulis beginning in the 1960s, holds that mitochondria are descended from ancient bacteria that entered into a stable relationship with other cells.

At first, this idea met with a great deal of resistance. Margulis recounted that her first major article on the origin of eukaryotic cells, published in 1967, had been rejected by numerous journals before being accepted by _the_ *_Journal of Theoretical Biology_*. Today, the bacterial origin of mitochondria is widely accepted: they possess their own DNA, a double membrane, and characteristics that preserve the memory of that ancient alliance. In other words, each of us carries within our cells the trace of an encounter that took place billions of years ago between different forms of life.

This is not a spiritual metaphor: it is the history of life. But it is a story that also possesses symbolic power. It reminds us that identity does not arise solely from separation. Sometimes it arises from cooperation; from the integration of what was originally other; from the ability to build unity without completely erasing difference.

This is what Alfred North Whitehead had intuited, in a philosophical sense, when he wrote: *"The many become one, and are increased by one.*" Whitehead, an English mathematician and philosopher, did not mean that differences dissolve into an indistinct mass. He meant to point to a movement: reality is not merely made up of things that are already complete, but of events in which many influences converge, giving rise to a new unity. Every organism, every gesture, every moment of consciousness can be viewed in this way: not as a closed block, but as a temporary and unrepeatable knot of relationships.

Our body, too, is a node of this kind. The cells that compose it renew themselves, molecules flow in and out, tissues change, and the environment leaves a continuous trace within us. And yet we are not merely a formless stream. There is a continuity, a living form that persists through change. The paradox lies precisely here: we are the same without ever being identical.

From this perspective, life is not a collection of components. A heart removed from the body is no longer a heart in the full sense of the word, even if it retains its anatomical structure for a certain time; an isolated cell does not account for an organism; a neuron, however thoroughly studied in minute detail, does not on its own contain a memory, a beloved face, or the meaning of a promise. The parts are real and indispensable. But the living being emerges only when— —the parts enter into a dynamic relationship, when they support, limit, and transform one another.

This is not a call to dismiss scientific analysis. Without the ability to distinguish, measure, and isolate, we would not have modern medicine, genetics, physiology, or molecular biology. The problem arises when this method—which is extremely useful for understanding the parts—is mistaken for a complete description of existence. Taking apart a violin can teach us a great deal about the wood, the strings, and the soundbox; but it does not allow us to hear a sonata. To hear it, the elements must once again become relationship, tension, vibration, and form in time.

Aldo Leopold, an American forester and thinker, formulated a similar idea in the field of ecology. In his famous **A Sand County* Almanac*, published in 1949, he proposed what he called an "ethics of the land": something is right, he wrote, when it tends to preserve "the integrity, stability, and beauty of the biotic community." The word "community" is important here. A forest is not merely a collection of tree trunks; a river is not merely flowing water; a landscape is not a neutral backdrop against which life unfolds. They are systems of interdependence, exchange, and fragile balances.

Leopold had also learned this lesson from firsthand experience. In 1909, as a young forest ranger in the Territory of New Mexico, he took part in the killing of a female wolf. Years later, he recalled seeing a "fierce green light" fade from her eyes—an image that became the centerpiece of his famous essay *"Thinking Like a Mountain.*" Over time, he came to understand that eliminating wolves—as many believed at the time would be beneficial for deer—did not simply mean removing a predator. It meant altering the entire rhythm of the mountain: the vegetation, the prey, the soil, and the very course of transformation.

The unity of life is therefore not merely a comforting abstract idea. It has practical consequences. If we think of ourselves as beings separate from the world, nature easily becomes a storehouse of resources. If we recognize ourselves as part of a web of life, the relationship changes: not because we suddenly become pure or innocent, but because we understand that causing unmitigated harm to our surroundings means damaging the very conditions that make our own existence possible.

Schrödinger, however, took the question even further beyond ecology. It was not enough for him to say that organisms are interdependent. He suspected that even the division between the knower and the known needed to be rethought. In this sense, he felt a kinship with certain currents in Indian philosophy, which he had encountered through translations and studies available in the first half of the twentieth century, as well as through the work of Arthur Schopenhauer. It is worth noting that Schrödinger had only a limited knowledge of Sanskrit and was not an academic specialist in Indian traditions. His interpretation was that of a European physicist-philosopher—intense but inevitably rooted in his own time.

What struck him most was the idea, present in the Upanishads and developed by "Advaita Vedānta," that behind the multiplicity of forms there lies a unified reality. The language of the Upanishads is different from that of science and should not be confused with it. When the Chāndogya Upaniṣad repeats the formula *"tat tvam* asi"—*"you are* that"—it is not stating a physical theory of the universe. It is proposing an inner exercise, a transformation of perspective: what we believe ourselves to be as a separate ego may not fully encompass who we are.

Here we are entering a symbolic and spiritual realm. We can listen to it without expecting it to be proof. We may even choose not to subscribe to it, while keeping the question " " open. The value of this encounter between Schrödinger and the East does not lie in fabricating a false "Vedantic physics," but in seeing how two distant languages can converge on the same unease: is the image of a world composed of isolated fragments truly adequate?

Spinoza, in the seventeenth century, had formulated the question in different terms, yet ones that are surprisingly close. For him, mind and body were not two distinct substances: they were two ways of expressing a single reality. *"The order and connection of ideas is the same as the order and connection of things*," he writes in *the* *Ethics*. We must not force the comparison: Spinoza is not Schrödinger, and Schrödinger is not a commentator on Spinoza. Yet both reject the temptation to imagine thought as a mysterious guest imprisoned within matter, or matter as a lifeless object to which, inexplicably, consciousness is added at some point.

Perhaps the most delicate point concerns the self itself. To say that individuality is in part an appearance does not mean to say that we do not exist. A person's pain is not illusory; their dignity is not a deception; their choices have consequences. Appearance, here, does not mean falsity. It means a partial way of seeing. It is true that we are individuals, but perhaps we are not merely individuals. We are unique forms of life, and at the same time we are permeated by what precedes and transcends us: our biological heritage, the language we have inherited, collective history, the air, food, our affections, and the matter of the stars from which the elements of our bodies originate.

This awareness can lead to two opposing outcomes. It can generate bewilderment: if I am not a perfectly separate entity, what remains of my identity? Or it can generate a more sober form of belonging: I am not the " " center of the cosmos, but neither am I a disconnected accident. My existence is irreducibly personal and, at the same time, inseparable from a broader fabric.

Perhaps this is the most fruitful lesson we can draw from Schrödinger, without attributing to him certainties he did not possess. Life no longer appears to us as an object placed before us, to be classified and dominated from the outside. It includes us. We are one of its momentary forms of consciousness; a point where matter, memory, relationship, and mystery find their voice.

We do not know if, behind our differences, there truly exists an ultimate unity. Science cannot declare it so, and philosophy cannot turn it into an indisputable fact. But we can acknowledge that absolute separation is too simplistic a fiction. We can look at the body, the ecosystem, another human being, and even our own consciousness with less possessive arrogance.

The self remains necessary: it is the name by which we take responsibility for our actions, preserve our memories, love someone, and face the fear of death. But perhaps the self is not a prison. Perhaps it is a window. And beyond that window lies not the void of an alien universe, but the inexhaustible depth of life, which, for a brief moment, gazes at itself through us.

Where does the body end, and where does thought begin?

Dualism, in its simplest form, is the idea that two radically different realities exist: on the one hand, matter—extended in space, measurable, subject to causes; on the other, the mind—invisible, inner, capable of feeling, choosing, and remembering. It is a division that seems almost inevitable. A stone falls, a cell divides, a neuron transmits an impulse: all of this belongs to the physical world. But the pain of loss, the red of a sunset, the sudden question about the meaning of one's own existence—how could these be reduced to the movement of particles?

Schrödinger never dismissed this difficulty with a brilliant formula. On the contrary, he considered it one of the open wounds in our worldview. The physicist who had given quantum mechanics one of its fundamental equations did not believe that science justified treating consciousness as a secondary detail, a sort of vapor produced by the brain. But neither did he consider it necessary to imagine an immaterial substance separate from the body, as if each of us were a mind imprisoned for a few decades within a biological machine.

His intuition, fueled as much by physics as by his reading of Indian philosophies, was both more radical and more cautious: matter and mind might be two ways in which the same reality manifests itself. Not two worlds awkwardly glued together, but a single world capable of appearing, from the outside, as a body and a physical process; from the inside, as experience.

This is not a conclusion of quantum physics. It is important to state this clearly. The equations of physics do not prove that the universe is conscious, nor do they prove that everything is mind. The transition from scientific description to a comprehensive view of reality remains an interpretive and philosophical one. However, twentieth-century physics has helped to undermine the old image of a universe made up of rigid, fully separate objects moving through empty space like marbles on a table.

Matter, as observed with the tools of contemporary physics, is not the opaque, inert mass that many past philosophies had imagined. An atom is not a tiny solar system; a particle is not always a localized ball; a living body is not merely an assembly of components. This does not mean that matter is an illusion. It means, more simply, that its nature is less straightforward than everyday experience suggests.

And perhaps even the mind, when viewed closely, is less separate from the world than we like to believe. In 1641, in his *Meditations on First Philosophy*, René Descartes articulated with extraordinary clarity the problem that still haunts us today. On the one hand, there is the *res extensa*, bodily reality, which occupies space and can be described geometrically; on the other, the *res cogitans*, thought, which doubts, imagines, and wills. The Cartesian distinction was not an abstract whim. It offered a powerful way to protect inner experience from being reduced to a mechanism and, at the same time, to leave to science the task of studying nature without depending on religious authority.

But that useful and fruitful move left a difficult question: if mind and body are so different, how can they interact? How can an inner decision cause a hand to raise? How can a physical wound be transformed into experienced pain? Descartes sought an answer in the pineal gland, a small structure at the center of the brain. Today we know that the pineal gland is not the bridge between an immaterial soul and the body. The question, however, has not disappeared: it has simply become more subtle.

For centuries, we have inherited—often without realizing it—the idea that we are divided into two. We say, "I have a body," as if the speaker were something distinct from the body they possess. We say, "My brain produced this emotion," as if there were a separate observer, sitting somewhere behind the eyes, receiving messages from a neural machine.

Yet our most ordinary experience tells a different story. We do not first encounter a body, and then a mind that inhabits it. We encounter hunger, fatigue, relief, desire, fear; we encounter the world through a posture, a breath, an embodied memory. Bad news doesn't just "hit" our consciousness: it knots our stomach, alters our sleep patterns, and tenses the muscles in our neck. A melody heard in childhood can resurface, decades later, not as an archived fact but as an entire emotional atmosphere: a room, a voice, an afternoon light.

The body is not a box containing inner life. It is the concrete way in which inner life participates in the world. Schrödinger addressed this crux above all in his most reflective writings, published after the years of the great revolutions in physics. In *"Mind and Matter,"* a series of lectures delivered at Cambridge in 1956 and published as a book in 1958, he observed that science, in order to obtain precise descriptions, had methodically separated the observer from the observed world. It had been an extraordinarily effective choice. To measure the trajectory of a planet or the chemical composition of a star, it was necessary to set aside emotions, intentions, and personal meaning.

But a methodological choice can, if we are not careful, turn into a comprehensive view of existence. We may end up believing that only what remains after excluding the subject is real: numbers, masses, fields, and measurable relationships. And then consciousness—which is the very condition by which the world appears to us—becomes an embarrassing element, almost a belated accident in an otherwise impersonal universe.

Schrödinger did not deny the legitimacy of scientific description. He knew its power all too well. But he believed it to be incomplete when it claimed to exhaust reality. In *"My View of the World,"* a text written in German in 1925 and published in expanded form in 1961, he formulated a phrase that has become famous: *"The total number of minds is one."* The statement may sound disconcerting, even excessive. It should not be read as an experimental result nor as an invitation to deny the differences between people. Schrödinger did not mean to say that individual sufferings are unreal, or that particular consciousnesses are interchangeable.

He was pointing to something else: the experience of being a separate self may not be the final word on our condition. It may be a perspective necessary for daily life, but not the definitive structure of reality.

Here his reflection draws close to the Upanishads and the non-dual tradition of India, which he had read carefully. In those ancient texts, the idea recurs that multiplicity is not false in the trivial sense of the word—rivers, trees, faces, and pains exist—but that it does not constitute a collection of isolated realities. The famous phrase *"tat tvam* asi" (*you are that*) from the " " appears in the Chāndogya Upanishad (probably composed between the 8th and 6th centuries BCE). In its context, it is not a consolatory slogan: it is an invitation to question the relationship between the individual self and the broader principle of reality.

Schrödinger found in this tradition a language capable of expressing what the European vocabulary tended to fragment: the relationship between subject and object, body and mind, individual and totality. For him, it was not a matter of replacing physics with Eastern spirituality. Rather, it was a matter of recognizing that one civilization can preserve questions and insights that another—focused on its own technical achievements—has sometimes left in the background.

A European thinker from a bygone era, Baruch Spinoza, had proposed something similar. Born in Amsterdam in 1632 into a Jewish family of Portuguese origin, Spinoza led a reclusive life that was often financially precarious; he even worked as a lens grinder at a time when optical instruments were expanding our understanding of the heavens and the microscopic world. In his major work, the *Ethics*, published posthumously in 1677, he argued that there are not two separate substances—mind and matter—but a single infinite reality, which he called *"Deus, sive* natura"—*"God, or Nature."*

This is a formulation that can easily be misunderstood. Spinoza's God is not a heavenly sovereign who decides, judges, and intervenes from the outside. It is the name of reality in its inexhaustible unity. Thought and extension—that is, what we call mind and what we call body—are, for Spinoza, two aspects of the same nature.

From this derives one of his most famous propositions: *"The order and connection of ideas is the same as the order and connection of* things" (*Ethics, Part II, Proposition 7*). This does not mean that every thought faithfully reproduces the world, nor that it is enough to think intensely about something for that thing to happen. It means that mind and body are not two separate realms forced to exchange mysterious messages. They are two ways of interpreting the same process.

Spinoza knew nothing of neurons, DNA, or quantum physics. Schrödinger, for his part, could not treat *the* *Ethics* as a scientific theory. Yet both sensed the inadequacy of the image that radically separates human interiority from nature. In this separation, nature becomes a silent backdrop, a reservoir of objects; the mind, a foreign guest observing the world without truly belonging to it.

Contemporary biology makes this sense of alienation even less convincing. We know that the human brain does not function as an isolated commander: it is immersed in an organism regulated by hormones, microbes, the immune system, sleep rhythms, social relationships, and the environment. Emotions cannot be confined to a specific point within the skull. They are configurations of the entire living being.

Neurobiologist Antonio Damasio, born in Lisbon in 1944, has long emphasized this point. In his studies of patients with brain injuries and in his books, including *Descartes'* Error* (1994), he has shown just how fragile the fantasy of a pure, disembodied reason is. Decisions do not arise from a cold calculation disrupted by emotions from the outside; feeling, bodily memory, and rational evaluation are intertwined. This does not solve the mystery of consciousness, but it makes it harder to think of the body as merely a vehicle for the mind.

In the early twentieth century, Jakob von Uexküll called the "world-environment" (*Umwelt*) specific to every living being: not the world in the abstract, but the world as it is encountered by a particular form of life . The idea does not imply that every species invents its own reality at will. In the example of the coin, the branch is real, the mammal is real, the smell is real. But what appears significant depends on the body that perceives it, on the capabilities it possesses, and on the kind of life it leads.

We, too, always live within an Umwelt. We do not see the world from anyone's perspective: we see it from here, through a human body, a language, a history, a nervous system, a culture. This does not lock us into a private prison; rather, it reminds us that knowing is always a relationship. There is not, on the one hand, a completely mute material world and, on the other, a consciousness that arbitrarily attributes meaning to it. There is a continuous encounter between an organism and its surroundings.

The French philosopher Maurice Merleau-Ponty, in the lectures collected posthumously in *Nature*, articulated this insight with memorable words: the body is not merely a thing among things, but *"our general means of having a world."* We do not stand before reality as spectators seated in a darkened room. We are already involved in it, exposed, vulnerable, capable of responding.

At this point, we must avoid two opposing shortcuts. The first is hasty materialism: everything would be merely matter, and sooner or later, conscious experience would be explained as a secondary effect, devoid of any substance of its own. The second is equally hasty spiritualism: matter would be a mere appearance, and consciousness could do without the body, natural history, and biological fragility.

Schrödinger does not lightly invite us in either of these two directions. He invites us to set aside the pretense that we have already understood what, at the core, both matter and mind truly are. The former is no longer merely the aggregate of solid objects imagined by common sense; the latter is not a phantom " " that can be separated without residue from bodily life. Both are terms that denote two facets of our experience of reality.

When a musician performs a piece, we can describe the vibrations of the strings, the movement of the air, and the brain activity of the listener. These are true and necessary descriptions. But on their own, they do not contain the meaning of that music: the tension of anticipation, the feeling of a form coming to fruition, the memory that a musical phrase awakens. Not because meaning violates the laws of nature, but because it is not of the same kind as a frequency or a mass.

Perhaps the limit does not lie in the world, but in the overly narrow categories by which we divide it. To say that matter and mind belong to a single reality does not mean erasing their differences. It means refusing to accept that difference must become absolute separation. The body remains a body; thought remains thought; each person's experience remains irreducibly situated. But the boundary between what we are and what we call the world may appear less like a wall and more like a threshold.

It is on that threshold that Schrödinger's thought retains its power. It offers neither a definitive proof nor easy consolation. It suggests a discipline of the gaze: do not treat consciousness as an embarrassing exception in nature, and do not treat nature as what remains once we have expelled every trace of experience from it. Perhaps we are not isolated minds observing an alien universe. Perhaps we are one of the forms through which the universe, for a brief moment, becomes aware of itself.

The self as an emergent phenomenon.

Among all the evidence of daily life, few seem more indisputable than the self. I remember, I desire, I fear, I choose; I feel my body as my own and spontaneously distinguish what is “inside me” from what is outside. Yet it is precisely this immediate certainty that deserves careful scrutiny.

This is not a matter of denying personal experience, nor of dissolving moral responsibility into a philosophical fog. It is, more simply, a matter of asking ourselves what we mean when we utter that very short yet enormous word: “I.” Do we mean a compact reality, always identical to itself, separate from the rest of the world? Or are we referring to a process: a continuity constructed by the body, memory, relationships, language, and consciousness?

Schrödinger was not a neuroscientist and could not have known about the experiments on the brain that today shed new light on this question. But his reflection on the unity of life led him to question the idea that every human being is a closed monad, a small inner fortress inhabited by a distinct owner. Physics, on its own, does not prove that the self is an illusion. However, it can undermine a certain naive image of the individual: that of an isolated, self-sufficient entity, absolutely separated from everything that generates it and permeates it.

The appearance of being separate.

When we say "I," we usually imagine something fairly simple: an inner center that possesses a body, receives impressions from the world, retains memories, and makes decisions. It is a powerful, almost inevitable figure. Language itself reinforces it: I think, I want, I'm hungry, I'm the same person I was yesterday. Grammar offers us a stable subject; social life—from a birth certificate to a signature at the bottom of a contract—gives it precise contours.

Yet our most basic experience tells a less linear story. We are not identical to ourselves in childhood, in illness, when falling in love, in mourning, or in old age. Our bodies, habits, fears, political beliefs, and even the way we remember what has happened to us change. And yet a sense of continuity persists: we recognize ourselves in the person we once were, even when we look back on that person with amazement or embarrassment.

Perhaps the self is not an immutable entity that resists change, but rather the fluid form through which change manages to become a story.

The Scottish philosopher David Hume, in his **Treatise of Human* Nature*, published between 1739 and 1740, articulated this insight with almost disarming candor. Searching within himself for a permanent core, he wrote:

"*When I look more closely into what I call myself, I always encounter one particular perception or another: heat or cold, light or shadow, love or hate, pain or pleasure. I can never grasp myself without a perception.*"

Hume did not mean to say that we do not exist. He meant to say that, in direct experience, we do not find a separate possessor of those experiences. We find thoughts, sensations, images, memories, desires. We find a succession, an interweaving, a flow. The idea of a stable self might be the way in which the mind gathers and connects this multiplicity.

A century and a half later, William James, one of the fathers of modern psychology, would express a similar idea in a memorable phrase: *"Thoughts themselves are the thinkers*." (The Principles of Psychology, 1890).

This statement should not be taken as a play on words. James wanted to free consciousness from the image of a small spectator sitting behind the eyes, separated from what he sees. There is not necessarily a distinct "someone" observing the mental flow from the outside; the flow itself, in its ability to connect, select, remember, and orient itself, produces the sense of a personal perspective.

Here it is important to be precise. To say that the self is an emergent phenomenon is not the same as saying that it is false, useless, or negligible. A melody, too, is emergent: it is not found in a single note, but that does not make it unreal. It truly exists; it can move us, it can be recognized, and it can change our mood. However, it is not something hidden behind the sounds: it is a form that emerges thanks to the ordered relationship between the sounds. Perhaps the self resembles a melody more than a stone.

It emerges from a bodily continuity, from an ever-partial memory, from a nervous system capable of integrating signals, from a language received from others, from affections, wounds, promises, and social roles. It is not an object that can be pinpointed in a corner of the brain. No neurosurgeon, upon opening a skull, could point to it with tweezers and say, "Here is the self." They would find networks of cells, electrical signals, chemicals, and regions that collaborate in various ways. The self is not a particle among the brain's particles; it is the dynamic profile that can arise from their coordinated activity within a living body, immersed in a human world.

This is the level of empirical research: not a final verdict on the nature of the person, but an important correction to the spontaneous image we have of ourselves.

A famous experiment demonstrated just how much our sense of the body is, at least in part, constructed. In 1998, researchers Matthew Botvinick and Jonathan Cohen published a study in *the* journal *Nature* that would become known as *the "rubber hand illusion*." A participant sat with one hand hidden from view; an artificial hand was placed in front of them. If the experimenter simultaneously stroked both the hidden real hand and the visible rubber hand, after a short time many participants reported a surprising sensation: the fake hand seemed to belong to them.

This does not mean, of course, that the body is imaginary. It means that the brain is constantly constructing a map of the body, integrating vision, touch, limb position, and expectations. Even the most immediate boundary—"this hand is me"—is not merely a raw fact: it is also the result of silent, ongoing coordination.

The bodily self, therefore, is not an immovable sovereign. It is a fragile and ceaseless achievement. Something similar emerges from studies of patients who, in the 1960s, underwent surgical separation of the corpus callosum—the large bundle of fibers connecting the two cerebral hemispheres. These procedures were performed in some severe cases of treatment-resistant epilepsy, in an attempt to limit the spread of seizures. Experiments conducted by Roger Sperry and, later, by Michael Gazzaniga showed that the two hemispheres could process information partially independently.

In one of the most well-known experimental scenarios, information presented only to the right hemisphere— , which, in most people, has more limited access to language—could influence behavior without the subject being able to verbally explain why. The left hemisphere, which is more adept at producing coherent narratives, sometimes tended to provide a plausible but fabricated explanation after the fact. Gazzaniga called this tendency "the interpreter": the narrative mechanism by which the mind unifies actions, impulses, and perceptions into a meaningful story.

We must not turn these studies into a caricature. People with split brains are not "two people in one body," and their cases do not justify simplistic conclusions about consciousness. But those experiments teach us something sober and decisive: the unity through which we feel like one is not a trivial given. It is an achievement. Our mind is constantly constructing a narrative of itself, and often does so with greater confidence than it has grounds for.

The British philosopher Derek Parfit, in his 1984 book **Reasons and* Persons*, brought this question into the realm of personal continuity. His formulation is deliberately provocative: "Personal identity is not what matters."

Parfit did not argue that people, our affections, or the consequences of our actions do not matter. On the contrary, he urged us to distinguish between the idea of a perfectly cohesive identity and what truly binds us over time: the continuity of memory, plans, responsibilities, and caring relationships. If we imagine the self as an indivisible block, any change may seem like a threat. If, on the other hand, we think of it as a living continuity, then we can better understand why we remain ourselves even as we change.

This perspective also sheds light on a truth we know long before we articulate it: we are individuals, but we are never merely individuals. Our name is given to us by others. We did not invent the language in which we say "I." Our earliest memories are often brought back to us through photographs, family stories, and repeated gestures. Even our image of ourselves is shaped by the gazes we receive: approval, rejection, friendship, humiliation, love. There is no human self that is born fully formed and only later enters into relationships. Relationships come first—or at least grow alongside what we call identity.

The sociologist and philosopher George Herbert Mead, who taught at the University of Chicago in the early decades of the 20th century, described the self as something that is formed through encounters with others. In *Mind, Self, and Society*, published posthumously in 1934, he wrote: *"The self is not present at birth, but is born in the process of experience and social activity.*"

It is a statement that retains an almost unsettling power. We are not merely creatures who possess relationships: we are creatures made, in an essential sense, of relationships. This does not erase our uniqueness; it explains it in a richer way. Each of us is unique not because we are separate from everything else, but because no one else occupies exactly that point of intersection between a body, a history, a constellation of bonds, a memory, and a horizon of possibilities.

Schrödinger would have recognized in this image a resonance with his own aversion to absolute individualism. In his philosophical reflections—which matured especially after the era of the great discoveries in quantum physics—the issue was not to eliminate the concrete person, with their joys and sufferings. It was to question the rigid boundary that separates one consciousness from another, as if each were a closed room.

However, an important distinction must be maintained. Quantum physics has not discovered that "we are all one mind," nor has it provided experimental proof of the ancient doctrines of India. When a physical theory describes correlated systems or challenges certain classical notions of separation, it is speaking of measurable phenomena and precise formalisms. The leap from there to a spiritual vision of unity is an interpretive leap, not a necessary deduction.

And yet it would be equally naive to ignore that certain scientific concepts, without proving metaphysical theses, can undermine the credibility of a certain mental habit: that of considering ourselves isolated by nature and connected to others only by chance.

At this point, Schrödinger's reflection approaches—though does not yet coincide with—one of the great insights we will encounter in the Upanishads and in "Advaita Vedānta." In that tradition, the separate self is not simply a lie to be exposed with contempt; it is the limited way in which consciousness experiences itself when it identifies solely with an individual history, with a body, with a bundle of memories and desires. The question is not whether this experience exists—it does, and it can be painfully real. The question is whether it exhausts who we are.

One of the oldest and most striking images comes from *the *Milindapañha**, a Buddhist text. In the dialogue, the monk Nāgasena responds to King Milinda. The king asks what a person is. Nāgasena then uses the example of a chariot: Are the wheels the chariot? Is the steering rod the chariot? Are the axle, the seat, and the reins the chariot? No single element, taken on its own, is the chariot; and yet, when those elements are arranged and function together, we call that whole "chariot."

The image does not say that the chariot does not exist. It says that it exists in a different way than we imagine solid, independent things to exist. It is a real, practical, recogniz ; but it depends on parts, relationships, and conditions.

Perhaps we can exercise the same caution when speaking of the self. It is not an illusion in the trivial sense of something that does not exist. It is a provisional form, a center of experience, a narrative and bodily continuity. But it becomes a source of confusion when we imagine it as an isolated substance, the absolute owner of the life that flows through it.

Viewing individuality in this way does not make us any less responsible. If anything, it might make us more attentive. If my "self" is not a fortress, then what happens to others is never completely external to my existence. If my identity has been made possible by countless connections—biological, familial, linguistic, historical, ecological—then autonomy does not equate to self-sufficiency. Rather, it means learning to respond, with clarity, within a narrative of which we are not the sole authors.

The self remains necessary: to love, to promise, to create, to ask for forgiveness, to take on a task. But perhaps it no longer needs to be idolized. It can be recognized for what it is: a threshold, not a prison; a singular voice, not a voice separated from the chorus of the living.

Life as a manifestation of the One.

At this point, we must be mindful of a crucial distinction. Physics does not prove that the universe is "One" in the spiritual or religious sense of the term. Its theories describe relationships, fields, transformations of energy and matter, and laws that operate equally in stars and cells. On their own, they do not provide us with a doctrine of the soul or proof of the absolute unity of all things.

Yet, for Erwin Schrödinger, the scientific worldview made the idea of totally separate individuals—each enclosed within its own boundaries—increasingly unconvincing. Life appeared to him as an event of continuity: matter that organizes itself, renews itself, and transmits itself; distinct forms that, though emerging for a limited time, remain immersed in a broader fabric.

When we speak of the One, therefore, we must not imagine a gigantic object hidden behind the world, nor a mysterious substance that erases differences. The One can be understood, more simply, as the name given to an intuition: what exists is not merely a collection of things placed side by side. It is also a network of interdependencies, exchanges, transformations, and mutual connections. Life appears individual because every organism possesses a form, a history, and boundaries; but no organism produces the conditions of its own existence on its own.

A tree is not merely a tree. It is sunlight transformed into plant matter; it is water drawn up from the roots; it is soil formed from ancient rocks and the remains of other living beings; it is the invisible work of fungi, bacteria, and insects. We, too—no matter how much our consciousness makes us feel unique and separate—are a passageway for elements that precede and transcend us. As we breathe, we take into our bodies atoms that have passed through countless forms. As we eat, we transform other life into our own. As we die, we return what we have called "I" for a few decades.

This is not poetry superimposed on biology from the outside. It is a basic—though often forgotten—description of our material condition. But it can become poetry, and perhaps even wisdom, when we understand that our dependence on the world is not a flaw to be corrected: it is the very way in which we exist.

Schrödinger felt this consequence with particular intensity. Trained in the great tradition of European physics, he never relinquished scientific rigor; yet he felt that the image of the human being as an isolated observer, standing before an alien nature, was too limited. In various writings collected after his death, especially in *My View of the World*, drafted in the 1920s and published in English in 1961, he returned to a thought he had encountered in Indian traditions, particularly in the Upanishads and Vedānta: the plurality of consciousnesses might be the way in which a deeper reality appears to itself from many points of view.

This is not a conclusion drawn from quantum equations. Schrödinger himself, at least in his more philosophical writings, operated on another level: that of interpretation and the search for a worldview capable of reconciling inner experience and scientific knowledge. To confuse these levels would be a mistake. But it would be an equally serious mistake to dismiss his openness toward the East as mere personal eccentricity. For him, it was a serious attempt to address a question that science—precisely because of its method—leaves open: What does it mean to be a conscious subject in a universe made of the same matter of which we are made?

The Upanishads—texts composed in India over the course of many centuries and initially transmitted orally—do not speak the language of physics. They do not propose repeatable experiments or hypotheses subject to laboratory verification. They speak through dialogues, dense formulations, cosmic images, and radical questions. But at the heart of many of them lies a recurring idea that continues to exert a singular power: behind the ever-changing multiplicity of forms lies a fundamental reality, Brahman; and the deepest core of individual experience, the Ātman, is not separate from that reality.

In *the Chāndogya Upanishad,* the teacher Uddālaka Āruṇi instructs his son Śvetaketu. To explain to him what cannot be seen but makes what we see possible, he uses familiar, concrete images. He asks him to dissolve some salt in water. The salt disappears from view, but the water retains its flavor throughout. It is a simple experiment, not a scientific demonstration: a metaphor to illustrate that what matters most may not be immediately visible.

At the end of a series of teachings, Uddālaka repeats to his son a phrase that has become famous: *"Tat tvam* asi" (*You are that*). (Chāndogya Upanishad, VI, 8, 7.)

That "that" does not refer to the psychological ego, our life story, our character, or our ambitions. It does not mean: you, with your tastes and fears, are the master of the universe. Rather, it means that the root of our existence is not separate from the root of existence in general. What lives within us is not a monad abandoned in a hostile world, but a local expression of a shared reality.

It is a thought that may seem dizzying—and even dangerous if reduced to a slogan. If everything is one, then do differences no longer matter? Would one person's pain be identical to another's? Would individual responsibilities be nullified? The great spiritual traditions of India have never provided a single, simple answer to these questions. But a more mature understanding of unity does not erase the world of differences: it places it within a broader horizon. The unique individual does not vanish; it merely ceases to be the whole of reality.

The *"Īśa Upanishad,"* a brief yet intensely profound text, expresses this position in words that also carry a clear ethical implication: *"He who sees all beings in the Self and the Self in all beings feels no aversion toward anyone."*

Here we are no longer merely faced with a thesis on the nature of reality. We are faced with a discipline of the gaze. If the other is not a fragment entirely foreign to me, if their vulnerability is not radically separable from my own, then indifference becomes harder to justify. Not because the universe imposes a morality on us from the outside, but because the absolute separation upon which we often base our indifference begins to show its cracks.

This insight is not unique to India. In distant times and places, various thinkers have sought words to name a unity that is not uniformity. Plotinus, born in Egypt around 204 and active in Rome in the third century CE, called the One the principle that cannot be reduced to one thing among things. In the Enneads, he wrote: *"The One is all things and is none of them."* (VI, 9, 11).

There is no need to force a direct historical link between Plotinus and the Upanishads; we do not need one. It is more interesting to recognize a resonance: when thought grapples with the fact that the world is manifold yet not scattered, it often encounters similar language. The One is not an object we can point to. It is the name of that through which things, while remaining distinct, do not plunge into a disconnected solitude.

Buddhism, too, while not accepting the idea of a permanent Self as presented by Vedānta, arrives at a similar conclusion on the level of experience: what we call a person is not an isolated substance, but a constantly changing interweaving. The Japanese master Dōgen, who lived between 1200 and 1253 and founded the Sōtō Zen school in Japan, wrote in *Genjōkōan*: "*To study the Way of the Buddha is to study the self. To study the self is to forget the self. To forget the self is to be realized in the encounter with all ten thousand things.*"

The "ten thousand things" is the traditional phrase used in the Far East to denote the infinite variety of the world. Forgetting the self, in this context, does not mean becoming passive or losing one's dignity. It means ceasing to treat the self as a besieged fortress, as the immovable center around which everything should revolve. It is an invitation to discover that we are already immersed in what we thought we were observing from a distance.

Schrödinger was not a spiritual master, and it would be inappropriate to turn him into a guru of physics. He was a modern man, with his own tensions, contradictions, and an often uneasy relationship with his own era. But it is precisely for this reason that his voice remains compelling. He did not seek in Indian philosophy an exotic embellishment to make European science more fascinating. He sought a vocabulary that would allow him to think without distorting experience.

In his thinking, the problem of unity did not arise from a desire to escape the concrete world, but rather from a l refusal to view the world as a sum of fragments without internal connection. Life, from this perspective, is not merely what happens within an organism bounded by a membrane or skin. It is a process that spans bodies, generations, ecosystems, languages, and memories. Every living being is distinct; none is separate.

We can then return to the initial question: in what sense is life a manifestation of the One? Not in the sense that a cell, a brain, or a forest proves the existence of a divine reality. Biology does not assert this. Nor is it in the sense that all differences are worthless illusions. Pain remains pain, injustice remains injustice, and every life retains its unique character.

Life manifests the One in a more subtle and perhaps more demanding sense: it constantly reminds us that separation is relative. Every boundary is real, but it is not absolute. Our skin distinguishes us from the world, but it does not cut us off from the air; our nervous system organizes a personal experience, but it does not create from scratch the language with which we think; our name identifies us, but it did not create us.

Perhaps the ancient intuition of the One does not ask us to abandon our individuality. It asks us to stop mistaking it for a self-sufficient possession. Schrödinger recognized in this a possibility for reconciliation: not a definitive answer to the mystery of consciousness and life, but a broader stance in the face of it.

What if life were truly this? Not a multitude of solitudes forced to meet from the outside, but a single reality that, for a brief stretch of its journey, learns to look at itself through different eyes.

VI. The Upanishads and "Advaita Vedānta."

"You are that":

Chāndogya Upaniṣad, VI, 8, 7, anonymous Vedic tradition.

This is the phrase with which the teacher Uddālaka Āruṇi addresses his son Śvetaketu. "That" refers to the profound reality from which all things arise, in which they dwell, and to which they return. The phrase is repeated nine times in the dialogue.

"I am Brahman":

Bṛhadāraṇyaka Upaniṣad, I, 4, 10. This is not the grandiose assertion of an individual self proclaiming itself divine. On the contrary, it is the dissolution of the absolute separation between the innermost self and the universal principle.

"All this, in truth, is Brahman. This Self is Brahman."

Māṇḍūkya Upaniṣad, 2; anonymous Vedic tradition. One of the most concise and radical formulations of non-duality: the world is not separate from ultimate reality, but is its expression or manifestation.

"No living being is ever born; there is no cause that can produce a living being."

Gauḍapāda, Māṇḍūkya Kārikā, III, 48. This is a deliberately extreme thesis, to be read within the context of his philosophy: it does not deny change as an everyday experience, but maintains that, at the deepest level, nothing is truly separate from the unity of Brahman.

There is a moment in Erwin Schrödinger's intellectual journey when readers accustomed to thinking of him solely as one of the fathers of quantum physics may feel a slight sense of disorientation. The great Austrian physicist, author of the wave equation and inventor of the famous thought experiment involving the cat, did not limit himself to questioning atoms, waves, and probabilities. He also probed consciousness. And in doing so, he turned his gaze far beyond Europe: to the Upanishads and the Indian tradition of "Advaita Vedānta."

We must clarify right away the ground on which we are standing. We are not speaking here of physics in the strict sense of the term. No quantum experiment proves the existence of a universal consciousness; no equation compels us to identify the individual mind with the whole. To say otherwise would be a compelling simplification, but one that is unfaithful to both science and Indian philosophy.

Instead, this is where the history of ideas comes in. It involves Schrödinger's personal and philosophical attempt to make sense of a fact he felt was difficult to circumvent: each of us experiences the world from within a consciousness, yet the plurality of consciousnesses remains a mysterious problem. We see many bodies, hear many voices, and encounter many life stories. But can we truly explain what it means that there are many centers of experience? Or is multiplicity, at least in part, the way in which a single reality presents itself through countless forms?

The Upanishads, composed in India between the first millennium BCE and the early centuries of our era, do not offer a scientific theory of the universe. They are texts of inner inquiry, often structured as dialogues between teachers and students, rulers and sages, fathers and sons.

Their tone is far removed from that of a modern treatise: it proceeds through images, questions, silences, and phrases for meditation. Yet, precisely for this reason, they have exerted a tenacious fascination on very different European thinkers: Arthur Schopenhauer, who kept a Latin translation of the Upanishads by his bedside; the philologist Friedrich Max Müller, who edited important editions of them at Oxford; and, later, Schrödinger.

The core of this tradition can be expressed in two Sanskrit words: *Ātman* and *Brahman*. Ātman is the self in its deepest sense: not our life story, not our character, not the sum of our memories and fears, but that which makes experience itself possible. Brahman is boundless reality, the principle from which all things derive their existence. "Advaita Vedānta"—literally, *"non-dual* Vedānta"—asserts that, at their root, Ātman and Brahman are not two separate realities.

This is the echo that Schrödinger recognized in his reflections on consciousness. Not because quantum physics had "revealed" the Upanishads to him, but because his scientific work had accustomed him to be wary of overly immediate images. The solid, compact, separate matter that everyday experience presents to us is not the matter that twentieth-century physics has learned to describe. Beneath the surface of things, we find relationships, fields, probabilities, and mathematical structures. Not an unreal world, of course, but a world less simple than it appears.

The Upanishads perform a similar act on another level. They invite us not to take separation as the final word. I and the world, subject and object, birth and death, internal and external: these are necessary distinctions in ordinary life. No one could cross a street or keep a promise without recognizing these concrete differences. But are they also ultimate boundaries? Or are they boundaries that help us navigate experience, without exhausting what experience itself is?

In the famous teaching of Uddālaka Āruṇi to his son Śvetaketu, found in the *"Chāndogya Upaniṣad,"* the teacher seeks to help his son intuit the invisible through everyday objects. He asks his son to dissolve some salt in water. The salt disappears from view, but it remains present: wherever the water is tasted, its flavor is detected. Thus, Uddālaka suggests, the deepest reality is not an object among objects. It cannot be grasped by hand nor isolated like a sample in a laboratory, and yet it permeates all that exists.

At the end of each explanation, the teacher utters the phrase destined to become one of the most famous in all of Indian culture: *"Tat tvam asi,"* "You are that." This does not mean: you, with your preferences and ambitions, are the master of the cosmos. It means almost the opposite: what we call "I" is not a small, isolated island in the vast ocean of the world. Its root belongs to the very same depth from which everything else emerges.

"Advaita Vedānta," in the form made famous by the philosopher Śaṅkara, who likely lived between the 8th and 9th centuries, developed this insight with extraordinary consistency. Śaṅkara traveled across the Indian subcontinent, commented on the Upanishads, the Bhagavadgītā, and the Brahmasūtras, and helped shape one of India's most influential philosophical traditions. His fundamental question was not very different from the one that, in other languages, haunts all philosophical inquiry: what remains when we strip away everything that changes?

The body changes, thoughts change, emotions come and go, and beliefs mature or crumble. Even the image we have of ourselves changes over time. Yet, as the Vedantic tradition observes, there is one element without which none of these changes could be recognized: the very fact that they appear to a consciousness. Not a "thing" hidden within us, not a measurable substance, but the openness through which every experience can occur.

At this point, the concept of Māyā emerges—one of the most misunderstood terms in Indian philosophy. Māyā is often hastily translated as "illusion," and this is where the misunderstanding arises: according to Advaita, the world would be false, a sort of worthless dream. But that is not the point. Physical pain does not become any less painful simply because we call it an "appearance"; an act of cruelty does not become harmless; the beauty of a face, a city, or a landscape is not erased.

Rather, māyā refers to the way in which we mistake a reality that is relative, changeable, and dependent for the absolute. It is the error of taking the form for the whole, the part for the whole, the name for what the name attempts to denote. A wave is truly a wave: it has a form, a force, a beginning, and an end. But it is not separate from the ocean. If we consider it merely as an isolated figure, we lose sight of its simplest and most obvious nature: it is water.

This image, while not a proof, helps us understand why Schrödinger might have felt drawn to the Vedantic idea of a non-multiple consciousness. In his essay *"Mind and Matter,"* published in 1958, he wrote that the plurality of minds is only apparent and that, in truth, there is but one mind. It is a dizzying statement, and it must be handled with caution.

It does not mean that people are interchangeable. It does not negate the uniqueness of bodies, memories, wounds, or responsibilities. It does not sanction a detached spirituality that, in the face of another's suffering, responds, "We are all one." If unity becomes a way of ignoring differences, then it betrays both the ethics and the depth of Advaita.

The issue is more nuanced. Schrödinger wondered whether consciousness can truly be counted in the same way that objects are counted. We can count ten chairs in a room because each chair occupies a distinct location. We can count ten people because their bodies are separated in space. But is consciousness—considered not as a brain function observable from the outside but as an experience lived from within—truly a reality that can be divided in the same way?

We do not have a conclusive scientific answer. Neuroscience describes the correlations between brain activity and experience with increasing precision, but the transition from the observed brain to lived consciousness remains one of the great open problems. The fact that a certain neural circuit is active can be measured; the fact that there is something we experience as “us”—a color seen, a sudden memory, the pain of a loss—belongs to a different order of questions.

Advaita does not solve the problem in the modern sense of the word. It does not produce data, does not construct verifiable models, and does not replace empirical research. It does, however, do something else: it challenges the idea that individual experience is a private fortress, definitively separated from all others. And it invites us to ask ourselves whether our habitual identity is not a useful, yet partial, way of describing who we are.

Perhaps this is where the encounter between Schrödinger and India still retains its power today. Not because it allows us to merge science and spirituality into a single reassuring term. Rather, because it compels us to respect the difference between these languages, without giving up on dialogue. Physics teaches us to be rigorous about what we can measure and predict. The Upanishads ask us not to confuse the measurable with everything that is real or meaningful. Philosophy, finally, keeps the question open: what do we mean when we say "I"?

Perhaps it is not necessary to choose too hastily between a universe made up of many isolated consciousnesses and a single consciousness that expresses itself in the many. We can pause in the space between these two images. We can recognize that, in everyday life, we are irreducible individuals, responsible for our words and actions; and at the same time admit that our separation may not be the ultimate truth about reality.

It is in this space, rather than in a definitive answer, that the Upanishads continue to speak to Schrödinger—and, through him, to us.

Atman and Brahman.

There are phrases that, as they traverse the centuries, retain a power almost too great to be immediately grasped. One of these is found in the *"Chāndogya Upanishad*": *"Tat tvam* asi"

Just three words. Yet, if we take them seriously, they call into question the habitual way in which each of us conceives of ourselves. Who is that "you"? And what does that "That"—which goes unnamed, as if it were too vast to be given a name—signify?

The Upanishads are not an orderly philosophical system in the modern sense of the term. They do not proceed through clear definitions, chapters, or linked arguments. They are texts born of meditation, of the dialogue between teacher and disciple, and of persistent questioning about death, desire, breath, and consciousness. They often speak in images: a seed, a flame, a river, a hawk, a chariot, a grain of salt dissolved in water. They do not intend to offer a theory of the world that we can observe from the outside. Rather, they ask us to change the vantage point from which we look.

At the heart of this inquiry, two words emerge: Ātman and Brahman. Ātman can be cautiously translated as "Self." But we must immediately avoid a misunderstanding. It is not the psychological self we speak of today: it is not character, temperament, the sum of memories, personal history, or the image we have constructed of ourselves. Nor is it simply the individual soul in the sense in which this word has been used in the European tradition.

The Ātman is that which, according to the Upanishads, endures through the vicissitudes of experience: that which experiences change without being reducible to it. Thoughts change, emotions come and go, the body ages, social roles shift, and even the idea we have of ourselves can crumble or transform. Yet there is something that accompanies all of this: the simple fact that there is experience, presence, and awareness.

In one of the most profound passages of the Bṛhadāraṇyaka Upanishad, we read:

"You cannot see the one who sees; you cannot hear the one who hears; you cannot think the one who thinks; you cannot know the one who knows." (III, 4, 2)

This statement is not a riddle. It points to a precise limit: everything we observe can become an object of our attention; but the one who observes cannot be grasped in the same way, because he is already present in every act of observation. We can describe the brain, record electrical activity, analyze a memory, or recount a biography. But the very experience of being present to all of this cannot easily be placed before us as just another object.

Here, however, we must point out the decisive shift. The Upanishads do not stop at the idea of an individual inner core. They assert something more radical: the Ātman, the deepest Self, is not separate from Brahman, the universal principle.

Brahman, too, is a difficult word to translate. It is not a personal god; it is not a deity with a face, a will, or a sacred biography. And it should not be confused with Brahmā, the creator god of the Hindu tradition, whose masculine name denotes a distinct figure. Brahman, in its neuter form, designates the fundamental reality: that from which everything arises, in which everything exists, and to which everything— —in some sense belongs. The *"Taittirīya* Upanishad" presents it with a sober and dizzying formula:

"From which, in truth, these beings are born; from which, once born, they live; and to which, upon dying, they return: seek to know that. That is Brahman." (III, 1, 1)

We are not yet in the language of physics, and we must not force the comparison. Brahman is not energy; it is not the quantum vacuum; it is not a universal physical field. Physics formulates verifiable models of the structures and behaviors of nature; the Upanishads speak of the foundation of experience and of being. Confusing the two planes would be an evocative but vague shortcut.

If anything, their convergence can occur on another level: in the question that both, in very different ways, make inevitable. What do we really mean when we say "I"? And to what extent is the separation between observer and observed world a complete description of reality?

The most famous scene—the one already described, which expresses the identity of Ātman and Brahman—takes place between a father, Uddālaka Āruṇi, and his son Śvetaketu. The story appears in the sixth section of the *"Chāndogya Upanishad."* Śvetaketu has returned home after twelve years of studying the sacred texts. He has learned a great deal, but his father notices that this knowledge has also given him a certain air of arrogance. So, rather than offering him an abstract lesson, he proposes a series of simple experiments.

In one of them, Uddālaka asks his son to dissolve some salt in a bowl of water and to return the following morning. The salt is no longer visible. The father asks him to taste the water at the surface, in the middle, and at the bottom. Everywhere Śvetaketu tastes the saltiness. The salt is not visible to the eye, but it is present throughout the water. Uddālaka then concludes:

"That subtle essence which you do not perceive—the entire universe is composed of it. It is the real. It is the Self. You are That, Śvetaketu." (VI, 13, 3)

The phrase, *"Tat tvam asi,*" appears nine times in the dialogue. It is not a motto to be repeated to feel important, nor is it a promise of individual greatness. On the contrary, it is a correction of the ego. It does not say, "You, with your merits, your ambitions, and your opinions, are the master of the universe." It says something far less comforting and far more demanding: what you believe yourself to be—this separate individual, defined by his boundaries, his fears, and his history—does not exhaust who you are.

The identity between Ātman and Brahman does not, therefore, mean that the individual, in his particularity, coincides with the totality in a simple and literal way. Śvetaketu remains Śvetaketu; Uddālaka remains Uddālaka; the world of differences does not magically disappear. The point is another: beneath the differences, or perhaps at their very heart, there is a single root of being.

The *"Bṛhadāraṇyaka* Upanishad" expresses this insight with another of the great phrases of the Vedantic tradition: *"Aham* brahmāsmi" (I am Brahman.) (I, 4, 10).

Here, too, the risk of misunderstanding is high. If the phrase is uttered by the egoic self, it can become a formula for self-aggrandizement. But its original meaning points in the opposite direction. It is not the personal self that proclaims itself to be the Absolute; rather, it is the discovery that the Absolute is not elsewhere, separate from lived experience, confined to a distant heaven, or entrusted exclusively to the authority of a ritual. This insight does not eliminate suffering, responsibility, relationships, or death. It does not sanction indifference toward the world. If every being participates in the same fundamental reality, the other is not merely a limit to my freedom or a tool for my own ends. The other becomes the place where separation is questioned. From this arises a possible ethical consequence: harming the other is never merely an act that concerns "someone else."

Centuries after the Upanishads, the philosopher Śaṅkara, who lived in southern India probably in the eighth century, would give this perspective a rigorous philosophical form in *"Advaita Vedānta."* Advaita literally means "not two": that is, not simply "one" in the numerical sense, but the absence of an ultimate separation between the Self and the principle of the whole. The multiplicity of experience is not denied as if it were nonexistent; rather, it is considered incapable of having the final say on reality.

It is important not to apply this formulation indiscriminately to all the Upanishads. These texts contain diverse voices, images that are not always easily reconciled, and tensions that cannot be reduced to a single doctrine. Śaṅkara's *Advaita* is a major interpretation of that heritage, not the only possible way to interpret it. But it is the interpretation that has most clearly made the idea of the identity between Ātman and Brahman famous.

The path of this idea to Europe was long and winding. In 1657, the Mughal prince Dārā Shikōh, son of Emperor Shāh Jahān, had a collection of about fifty Upanishads translated into Persian, convinced that those texts concealed a point of convergence between Sufi Islam and Indian wisdom. He called the collection *"Sirr-i* Akbar" (*The Great Secret*). At the end of the 18th century, the French philologist Anquetil-Duperron translated that Persian version into Latin; the work, published between 1801 and 1802 under the title *"Oupnek'hat,"* found its way into the hands of Arthur Schopenhauer.

Schopenhauer was deeply impressed by it. Through an indirect, imperfect, and often difficult-to-read translation, he found in ancient India confirmation of his critique of individualism and his intuition of a common foundation underlying all beings. Later, in a Europe by then transformed by modern science, those same ideas would also captivate Erwin Schrödinger.

Schrödinger did not find scientific proof for quantum theory in the Upanishads. His equation, his research on wave mechanics, and the cat paradox belong to the realm of physics, with its own methods and verifications. But he found in the Vedantic tradition a language capable of accommodating a question that physics, on its own, cannot resolve: how can we conceive of the unity of conscious experience without reducing it to a sum of isolated centers?

For this reason, the Ātman and Brahman are not, in our journey, an exotic embellishment added to contemporary physics. They are a philosophical challenge. They compel us to distinguish between the person we are—unique, situated, vulnerable, responsible—and that deeper dimension that the Upanishads refer to as the Self.

We can accept or reject this vision. We can interpret it as metaphysics, as a spiritual exercise, as a symbol of radical solidarity, or as an ancient attempt to speak of the mystery of consciousness. But we can hardly engage with it without one question remaining open: if the boundary between "me" and "everything else" were not as definitive as it appears in everyday experience, what would change in the way we inhabit the world?

Māyā and Appearance.

Few words have met a more curious fate than Māyā. Having entered the Western lexicon through translations, novels, spiritual movements, and textbook simplifications, it has often become a sort of ready-made formula: "The world is an illusion." A phrase that may seem profound for a moment and then become dangerously convenient. If everything is an illusion, why take pain seriously? Why care for the sick, stand up against injustice, nurture a relationship, or take responsibility for one's actions?

"Advaita Vedānta" does not ask us at all to conclude that hunger is imaginary for those who suffer from it, or that a loss leaves no wounds because it belongs to the world of appearances. To confuse Māyā with a denial of the world is to miss the essential point. The phenomenal world—the world that appears in its colors, its bodies, its events, its conflicts—is not nothing. It is the concrete plane in which we live. But, according to this perspective, it is neither the whole reality nor its definitive form.

Let's proceed with caution. When we say that something "appears," in common parlance we often mean that it may not be true. But appearance is not always a lie. A sunset appears red; this does not mean the sunset does not exist. Rather, its color depends on specific physical conditions: sunlight passes through the Earth's atmosphere, air molecules scatter the longer and shorter wavelengths more effectively, and it is primarily the warm hues that reach our eyes. The red of the sunset is real as an experience and as a phenomenon, but it is not a simple, isolated, and immutable property of the object "Sun."

Here, physics has the final say: what we perceive arises from the interplay between an external reality, the material conditions of observation, and the structure of our senses. It does not follow that the world is a private dream. It follows, more soberly, that the way the world appears to us never completely coincides with what it is independently of us.

The Indian reflection on Māyā operates on a different, broader, and more radical plane. It does not merely remind us that the senses can be mistaken. It asks us whether the entire universe of distinctions—self and other, internal and external, birth and death, success and failure—has an ultimate substance or is a changing manifestation of a deeper reality.

The word māyā is ancient. It appears as early as the *Ṛgveda*, the oldest collection of hymns in the Vedic tradition, composed in northwestern India during the second millennium B.C. In those texts, it does not yet clearly refer to the "illusory world" of which Advaita will speak. It can mean extraordinary power, creative art, the ability to take on forms, or an enigmatic force. It is, so to speak, the power through which the divine reveals itself in manifold ways.

In a hymn of the Ṛgveda, the god Indra is celebrated for his māyā—his powers and transformations. We are therefore not faced with the thesis that what appears does not exist; rather, we are faced with the perception that reality possesses a fluid, creative character that is difficult to confine within a single definition.

This nuance remains important even in the later Upanishads. The *"Śvetāśvatara Upanishad,"* a text likely composed in the last few centuries before the Christian era, offers a famous passage:

"Let nature be recognized as Māyā,
and the great Lord as the one who possesses Māyā.
From that which is part of him, this entire world is pervaded."
(IV, 10)

This is a verse that calls for caution, because it belongs to a tradition in which the emphasis on the Lord, Maheśvara, is more pronounced than in classical Advaita. Nevertheless, it preserves a decisive insight: Māyā is not merely human error; it is the name of a power of manifestation. The world of forms is not a meaningless accident, nor a trap set by a deceitful god. It is the way in which unity presents itself in experience as multiplicity.

Centuries later, Śaṅkara would give this insight a formulation of extraordinary precision. His school, "Advaita Vedānta," asserts that Brahman, the fundamental reality, is non-dual: it is not divided into separate and independent parts. Our ordinary world, on the other hand, is the realm of differences. Here, everything seems distinct: this body is not that tree, this person is not that person, the past is not the future, desire is not its object. The error does not lie in perceiving differences. Differences do indeed operate in daily life. The error lies in believing that they are absolute, that they explain everything, that they are the ultimate structure of reality.

To clarify this point, Advaita commentators have often used the example of a rope mistaken for a snake at twilight. Let's imagine a dimly lit path. On the ground we see an elongated shape, and for a moment we step back: it looks like a snake. Our heart races, our body reacts, and the fear is genuine. Then someone shines a light on it, and we discover it was just a rope.

What happened? The snake wasn't there, but the experience of fear wasn't fake. It had real consequences. The rope was real, but it had been misinterpreted. The appearance wasn't nothing: it was a confused perception, superimposed on what was right before our eyes.

Śaṅkara opens his famous commentary on *the "Brahma* Sūtras" precisely with the idea of this superimposition: we attribute the characteristics of one thing to another. To the body, the mind, memories, and social roles, we attribute the character of a stable and separate "I"; to the Self, on the contrary, we attribute the limitations and vicissitudes of that which changes. It is an error that does not occur only in theory: it is the spontaneous gesture with which each of us says "I am this," "I am only my story," "I am identical with what I fear losing."

From this perspective, māyā is not a veil drawn over our eyes by some hostile power. It is the very way in which ordinary experience organizes the manifold and, at the same time, forgets its dependence on a unified ground.

Advaita uses an important word in this regard: *mithyā*. Translating it simply as "illusion" is misleading. Mithyā does not refer to something that is completely nonexistent, such as the child of a barren woman or a square circle. It refers to something that possesses a dependent, conditioned, relative reality. A clay jug is used to carry water; it can break, it can be bought or given as a gift: on the level of concrete life, it is real. But its form does not exist on its own; it depends on the clay, the potter's work, the firing heat, the use we make of it, and the time that will wear it down.

The jug is not unreal. Yet it is not a reality independent from beginning to end. Its identity is provisional, interwoven with relationships and transformations.

Perhaps we can recognize here a truth that does not belong solely to ancient India. Our personal identity, too, is concrete and yet unstable. We have a name, a face, a mother tongue, an age, a memory. All of this matters. But all it takes is a bereavement, an illness, a love affair, a migration, a war, or an unexpected discovery to show us just how much less solid the image we had of ourselves was than we believed.

In 1927, the writer Luigi Pirandello published *One, No One, and One Hundred Thousand*. The protagonist, Vitangelo Moscarda, begins with a seemingly minor observation made by his wife: his nose tilts slightly to the right. Moscarda had never noticed it. From that crack, an abyss opens up: if others see a man different from the one he believes himself to be, how many versions of him exist? Just one? One hundred thousand? None that are definitive?

Pirandello is not a Vedantic author, and it would be contrived to turn him into a commentator on Śaṅkara. But his novel helps us sense, within a context close to European sensibilities, the fragility of the self that we imagine to be solid. Advaita pushes the question even further: if the person we defend so stubbornly is made up of changing elements, where do we seek that which does not change?

Here we enter the philosophical realm—and, for those who find it possible, the metaphysical one. Advaita's answer is not that the world must be abandoned, nor that attachments are a mistake. It is that the world must be lived in without turning it into an absolute. One can love without claiming to possess. One can act with seriousness without believing that the result defines us forever. One can suffer without adding to the pain the idea that it is the whole truth of who we are.

This is not an invitation to easy serenity. On the contrary, recognizing the changing nature of things can be unsettling. Our most deeply held certainties often depend on the hope that someone or something will remain unchanged: the body, prestige, a relationship, a homeland, or a self-image. According to Advaita, Māyā begins to operate when we mistake these necessary but temporary forms for an unshakable foundation.

At this point, it is helpful to clear up a common misunderstanding. Quantum physics does not prove Māyā. It does not prove that matter is imaginary, nor that individual thought fabricates the world at will. Quantum phenomena show that, in the realm of the infinitely small, our spontaneous images of objects—which we always assume to have definite properties—encounter profound limitations. But to conclude from this that "everything is an illusion" is a leap that physics does not authorize.

When a laboratory measures a particle, when a detector registers a signal, when many researchers can verify a result, we are on the ground of science: public observations, instruments, calculations, and testable hypotheses. Māyā belongs to a different order of discourse. It is not an alternative physical theory; it is a question about the meaning of the reality we experience and about the relationship between its changing forms and their possible foundation.

Erwin Schrödinger, who was well aware of the difference between a scientific theory and a philosophical vision, did not seek in the Upanishads a shortcut to explain the experiments of quantum mechanics. His interest was more personal and broader: the difficulty of conceiving of a reality fragmented into a multitude of completely separate observers. Indian philosophy offered him a vocabulary with which to question that separation—not a scientific certificate to abolish it.

Māyā, then, does not tell us, "Nothing matters." Perhaps it says the opposite: precisely because everything is fragile, dependent, and transitory, everything deserves a less distracted gaze. The world is no less precious simply because we do not possess it ful . A falling leaf is not an illusion; it is a real, fleeting, unrepeatable event. Its transience does not strip it of meaning. If anything, it reminds us that reality does not coincide with what we are able to hold onto.

One question remains, and there's no need to resolve it too quickly. If what we call reality is a web of manifestations, relationships, and forms that arise and vanish, can we truly continue to regard ourselves as isolated entities, enclosed within absolute boundaries? Or is our daily experience of separation merely the first level of a broader narrative, one that never ceases to appear precisely because, perhaps, it has never ceased to conceal something?

Non-multiple consciousness.

Perhaps the most difficult idea to accept in "Advaita Vedānta" does not concern the external world, nor the changing nature of things. It concerns us. Or rather: it concerns the one within each of us who says "I."

We are accustomed to thinking of consciousness as a private matter. I feel my pain, I remember my childhood, I fear my death; you feel your pain, you hold onto other memories, you fear a different end. Between my inner experience and yours, there seems to be an insurmountable distance. We can tell each other what we feel, observe one another, love one another, hurt one another; but no one can directly enter into the consciousness of another.

This experience of separateness is so immediate that it seems indisputable to us. Yet "Advaita Vedānta" poses a question that cannot be easily dismissed: Are we truly many separate consciousnesses, or does multiplicity pertain to bodies, minds, memories, and points of view, while consciousness, at its most essential core, is one?

It is not a psychological question in the modern sense of the term. It does not ask whether human beings experience similar emotions, nor whether there is a collective mind made up of the sum of individual minds. It does not suggest that we all think the same things or that, by closing our eyes long enough, we could read the thoughts of a stranger. Advaita speaks neither of telepathy nor of a vast mental network that absorbs people into an indistinct mass spiritualism.

Its thesis is more understated and, precisely for this reason, more mind-boggling. There are many minds: they have different memories, different languages, different wounds, and different abilities. But the consciousness through which each mind appears to itself may not be numerically multiple. It might be like a single light that, passing through many windows, takes on different colors and shapes without actually dividing itself.

The image of light should not be taken as a proof. It is a metaphor, not evidence. But it can help us intuit what words struggle to express. In a large room, ten mirrors reflect the same lamp. Each shows a different image: brighter or dimmer, distorted or clear, partially covered in dust or cracked. It would be absurd to say that there are ten lamps simply because we see ten reflections. Yet it would be equally absurd to deny that the images are different.

Advaita proposes something similar: the person is a unique center of experience, but this does not mean that the consciousness that makes every experience possible must be an individual possession. There are many reflections; the light, perhaps, is one. In one of the great Upanishads, the *"Bṛhadāraṇyaka* Upanishad," we find a passage of almost disconcerting radicalism. The sage Yājñavalkya speaks to his wife Maitreyī and, through a series of questions, leads her to the very limits of language. At one point he states:

"Where there is duality, there one sees another,

one hears another, one thinks of another, one knows another.

But where everything has become the Self, with what could one see whom?"

The text belongs to an era far removed from our own. Yet its question retains its full force. If ultimate reality is undivided, what does it mean to know? Who knows whom? And what remains of the boundary between observer and observed?

It is important not to turn these phrases into a vague formula. The Upanishad does not deny that, in ordinary life, I see another person, hear a voice different from my own, and recognize a face. That difference is concrete and must be respected. Rather, the text asks us to consider whether the relationship between two people exhausts all that they are. It invites us to suspect that separation, though real in everyday experience, is not the final word.

For modern Western thought, shaped by the idea of the autonomous individual, this perspective may seem almost intolerable. Our civilization has painstakingly won the value of the individual: freedom of conscience, individual rights, responsibility before the law, and respect for difference. These are not secondary achievements, nor should Advaita be used to call them into question.

To say that consciousness is one does not mean that people do not matter. If anything, when taken seriously, this idea makes it harder to treat the other as a thing. If what lives in the other is not radically foreign to what lives in me, indifference is no longer merely a moral failing: it becomes a form of blindness.

But here we enter the realm of existential interpretation, not that of a proof. The Upanishads do not provide us with a political program, nor do they automatically resolve human conflicts. History teaches us that even the loftiest ideas can coexist with unjust societies. No philosophy, on its own, makes us good. What it can do is alter the nature of the questions through which we view ourselves and others.

Erwin Schrödinger grappled with this question in a personal and persistent way. He was not the first European to turn to Indian texts, nor was he the most systematic interpreter of Advaita. But he was one of the very few physicists of the twentieth century to openly declare how much that vision had meant to him.

Born in Vienna in 1887 and died in Vienna in 1961, Schrödinger spent pivotal years of his career in Zurich, Berlin, Oxford, and Dublin. It was in Dublin, where he directed the School of Theoretical Physics at the Dublin Institute for Advanced Studies beginning in 1940, that he developed many of the philosophical reflections later collected in his books and essays. These were not mere marginal notes by a great scientist distracted from his discipline. They were part of a quest that accompanied him until the end: how to conceive of life, the mind, and the unity of nature without reducing them to a mechanism devoid of inner life?

In his book *Mind and Matter*, Schrödinger formulated the famous statement: *"There is evidently only one alternative: the unification of minds or consciousnesses. Their multiplicity is only apparent; in truth, there is only one mind."*

It is a bold statement, even reckless if we read it as a scientific assertion. And we should not do so. Schrödinger's physics—that of the wave equation, quantum mechanics, particles, and atoms—does not prove that a single consciousness exists. No laboratory experiment has established that all subjective experiences are expressions of a single universal mind.

Here, physics in its strict sense is not speaking. Here, Schrödinger the philosopher is speaking—nourished by the Upanishads and troubled by the difficulty of explaining how consciousness can emerge from matter described only from the outside. It is essential to distinguish between these levels. His intuition may be philosophically fruitful; it does not, however, become a result of quantum mechanics.

Schrödinger did not even think that the individual was a mere, insignificant invention. He knew full well that human life unfolds in plurality: each person is born into a body, speaks a language, and experiences events that no one else experiences in the same way. What seemed questionable to him was the idea that these differences compelled us to imagine so many absolutely separate consciousnesses, like so many isolated islands in the ocean.

In another passage from *Mind and Matter*, he observes that the world is given to us only once: not once for each individual, as if there were private, incommunicable universes, but just once, in a form that appears manifold. This is a formulation that may seem obscure, but it touches on a fundamental fact. We never encounter "consciousness" as an object among objects. We encounter things, people, landscapes, memories, thoughts. And we encounter all of this in the very fact of being conscious.

Consciousness is that through which everything becomes present; this is why it is difficult to place it before us as we do with a stone or a neuron. We can study the brain with increasingly sophisticated tools. We can observe the areas that become active when we see a face, hear a melody, recall a word, or make a decision. This research is indispensable and has transformed our understanding of the mind.

And yet one question remains, often posed by the Australian philosopher David Chalmers in contemporary debate: why are certain physical processes in the brain accompanied by lived experience? Why is neural activity not merely a sequence of electrical and chemical exchanges, but manifests itself as the blue of the sky, the taste of coffee, the pain of a wound, or the longing for a voice?

We do not yet know how to answer this conclusively. Saying this is not a failure of science: it is an act of precision. Neuroscience has established increasingly solid correlations between brain states and conscious experiences. Damage to certain regions of the brain can alter memory, language, perception, and personality. Anesthesia can interrupt or alter consciousness; sleep transforms it; neurodegenerative diseases can painfully change its profile. Any serious discussion of consciousness must reckon with these facts.

But a correlation, however close, is not yet equivalent to a definitive explanation. The fact that a violin is indispensable for a sonata to resonate does not, by itself, tell us that music is identical to the wood, the strings, and the vibrations of the air. The analogy has obvious limitations: the brain is not a mere instrument separate from a hidden musician. Yet it helps us avoid confusing two distinct issues. One concerns the way in which experience depends on the body and the brain. The other concerns what that experience is, in and of itself.

Advaita addresses the second question with an answer that, for many, will be too challenging: consciousness is not a local product generated by the brain, but rather the principle through which the brain, the body, and the world themselves become knowable. The brain does not create consciousness out of nothing; rather, it makes possible an individual, limited, situated manifestation of it.

This is a metaphysical proposition. It should not be presented as an empirical fact. It is a view of the relationship between mind and matter, and it can be discussed, criticized, or rejected. But to dismiss it out of hand would be equally frivolous. It arises from an attempt to fully grasp a fact that we all know but that no one can observe from the outside: the fact that experience exists.

In the early twentieth century, the American psychologist and philosopher William James defended an almost opposite conception. In his book **A Pluralistic*

Universe*, published * * in 1909, James insisted on the plural, unfinished, and open nature of reality. He was wary of grand philosophies of the One when they risked absorbing concrete differences into a perfect unity indifferent to the dramas of history.

His objection remains valuable. If unity becomes a way of ignoring a person's particular pain, then it is a bad kind of unity. If it serves to tell those who suffer that their suffering does not matter because "everything is one," then it has betrayed its very meaning. A non-pluralistic consciousness, if it has any human value, should not erase singularity; it should make it even more inviolable.

An unexpected parallel comes from Romain Rolland, a French writer and 1915 Nobel Prize laureate in Literature. In a letter sent to Sigmund Freud on December 5, 1927, Rolland spoke of an "oceanic feeling": the sensation of a bond with the whole that has no clear boundaries. Freud took up the expression in **Civilization and Its* Discontents*, published in 1930, but stated that he did not recognize that experience in himself. He interpreted it cautiously, as a possible remnant of a very early phase of psychic life, prior to the clear distinction between the ego and the world.

It is an illuminating disagreement. Rolland perceives in human experience an openness toward the unlimited; Freud seeks an explanation for it in the history of the psyche. Neither can simply dismiss the other. The former reminds us that consciousness can transcend the usual image of the individual closed in on himself; the latter reminds us that inner experiences, even the most intense ones, do not interpret themselves.

Perhaps this is the most fruitful stance to take toward the Vedantic thesis of the one consciousness. We should not immediately ask ourselves whether it is "proven" or "absurd." We can first ask ourselves what it challenges. It challenges the idea that the biographical self is the definitive center of the universe. It challenges— —the conviction that the boundary of our skin coincides with the boundary of who we are. Finally, it challenges the habit of thinking of consciousness as private property, akin to a treasure kept in a locked room.

Yet it does not ask us to stop being people. If anything, it asks us to become people with less rigidity. We continue to call each other by name, to keep promises, to acknowledge responsibilities and differences. But perhaps we can do so without turning our identity into a fortress.

Non-multiple consciousness is not an answer that makes the mystery any smaller. It makes it more demanding. If what appears within me and what appears within another living being participates, in some way, in the same light of awareness, then every encounter takes on new significance. Not because distances disappear, but because distance may not be the ultimate truth.

VII. The observer is not outside the scene.

"It was not possible to formulate the laws of quantum mechanics in a fully consistent manner without referring to consciousness."

Eugene P. Wigner, "Remarks on the Mind-Body Question," in Symmetries and Reflections (1967).

"Consciousness is what makes the mind-body problem truly difficult."

Thomas Nagel, "What Is It Like to Be a Bat?" (1974).

"Physics is not about how nature is, but about what we can say about nature."

Niels Bohr, quoted by Aage Petersen, "The Philosophy of Niels Bohr," Bulletin of the Atomic Scientists (1963).

Discussing consciousness in relation to quantum physics requires a preliminary discipline: not to turn a scientific difficulty into a test of spiritual doctrines. Quantum theory does not prove that the universe is a mind, nor does it prove that each of us participates in a universal consciousness. Yet it has forced physics to take one fact seriously: every description of the world begins with an encounter between something that happens and someone—or something—that registers it.

For Erwin Schrödinger, this point was not confined to the laboratory. The problem of observation seemed to him to lead to a broader question: Who, after all, is the knowing subject? His answer was radical. According to Schrödinger, the idea that there are many separate consciousnesses is part of the way we normally experience the world. In everyday life, in fact, each of us perceives ourselves as a self distinct from others, with our own thoughts, sensations, and memories. However, at a deeper level, this separation is merely apparent: at the root of experience lies a single consciousness, which manifests itself through a multiplicity of individuals and points of view. This is a proposition that does not belong to experimental science in the strict sense of the term. It is a philosophical—and, in many respects, metaphysical—vision, informed by his reading of the Upanishads and "Advaita Vedānta." Precisely for this reason, Schrödinger remains a singular figure: a leading physicist who refused to completely separate the problem of matter from the mystery of the one who experiences it.

When we speak of an "observer" in quantum physics, we immediately risk a misunderstanding. The word suggests a human being standing motionless before an experiment—perhaps a physicist in a white lab coat watching a needle move across a dial. But in scientific practice, an observation can be made by a r photographic screen, an electronic detector, a cloud chamber, or a Geiger counter. A human eye does not need to witness the event for an atom to leave a trace.

It is important to state this unambiguously: quantum mechanics, in its operational form, does not require a human mind to magically produce physical facts. A photon can be absorbed by a plate, a particle can strike a detector, an atom can decay in an empty room. The idea that "consciousness creates reality" is not a necessary conclusion of quantum physics; it is a seductive but overly simplistic generalization.

And yet the problem does not disappear. For even when we entrust the measurement to a machine, a question remains: what does it mean for a result to have become a fact? The machine records a signal; another device can store that signal; a computer can process it. But the meaning of a recording—its appearing as knowledge of something—seems to bring us back, ultimately, to the realm of experience. A number on a screen is not merely a configuration of pixels: it is a number that has been read, understood, and placed in relation to a question.

Niels Bohr, who was among the founders of the most influential interpretation of quantum theory (the Copenhagen interpretation), insisted on a more sober and less spectacular point. For him, physics does not describe a reality observed from nowhere, as if we could place ourselves outside the universe and view its complete picture. It describes what we can say about nature based on communicable experiments. This is not a technical detail. It means that scientific knowledge is not a purely passive mirror: it is an ordered activity, composed of instruments, language, concepts, and public discourse.

Schrödinger shared the awareness that there is no description completely separate from the subject who knows it. But he went further. It was not enough for him to assert that the subject participates in the construction of knowledge; he wanted to ask himself what this subject is. And his answer, which matured over the years and was made explicit above all in the essays collected in **Mind and Matter**, was disarming: consciousness, strictly speaking, never presents itself as a plurality.

"*Consciousness is never experienced in the plural, but only in the singular*," he wrote. It is a brief, almost provocative statement. It seems obviously false to us, because we live surrounded by other people. We encounter other gazes, hear other voices, see other bodies suffering or rejoicing. We know, by analogy and from experience, that others are not inert objects. They have an inner life. They can surprise us, contradict us, love us, or hurt us. How, then, could consciousness be singular?

Schrödinger did not intend to deny the concrete existence of people, nor to propose an absolute solitude in which only one's own self would be real. That would be a form of solipsism—the thesis that only my mind exists and everything else is a representation of it. Schrödinger's position, if anything, goes in the opposite direction: not a single real person and all others unreal, but a single source of experience that manifests itself through many real people.

The distinction is crucial. Solipsism reduces the world to the individual self; Schrödinger, on the other hand, seeks to relativize the individual self. What we call "I" would not be the ultimate owner of consciousness, but rather a temporary and limited perspective of it: a biography, a memory, a body, a character—a particular window open onto the world.

The image of many windows can help us, provided it is not taken as proof. From a house overlooking a square, each window shows a different, portion of the same scene. One looks at the fountain, another at the bell tower, yet another at the people crossing the street. The images do not coincide: each depends on position, light, distance, and even the transparency of the glass. Yet this does not mean there are many squares. There is one square that appears in many ways.

Here, however, the metaphor breaks down. Consciousnesses are not passive windows. They are lives that feel, interpret, decide, and remember. And the hardest thing to conceive is precisely this: if there is only one consciousness, how are the radical diversity and inaccessibility of others' experiences possible? We do not feel another's pain in our own bodies. Another's joy is never identical to our own. The unity Schrödinger speaks of does not erase this distance. It seeks to place it on a different plane.

We can put it this way: on the human and practical level, we are irreducibly many. We have different names, responsibilities, bodily boundaries, histories, and wounds. Law, ethics, friendship, and politics all stem from this real plurality. But Schrödinger wondered whether this plurality was the final word on consciousness, or whether it was instead the way in which the single experience of the world presents itself as it passes through distinct organisms.

In this question, one can sense the influence of the Upanishads. In the Indian texts that Schrödinger read and engaged with for decades, the ātman—the deepest self—does not simply coincide with the psychological "I." It is not the sum of memories, preferences, and fears. It is that which, beneath these differences, is conceived as identical to Brahman, the universal principle. According to the Advaita, or "non-dual," tradition, the separation between subject and world is not entirely false—it has a concrete force in ordinary life—but it does not exhaust reality.

Schrödinger did not merely borrow Eastern terminology to cloak modern physics in mystery. That was not his style.

If anything, he found in those traditions an ancient expression of a concern that physics, on its own, could not resolve: how can the world be described as a multiplicity of objects, while experience always presents itself as a unified whole?

Every one of our perceptions is, in fact, unified. We can listen to music, see the color of a wall, feel fatigue in our body, and recall a face all at the same time; yet we do not experience these elements as fragments tossed into a sack. We experience them as "my experience right now." Neurology can show that many areas of the brain are involved in this integration. It can study the circuits of attention, memory, and perception. But the question remains as to how processes distributed throughout the brain become the inner face of experience: the red we see, the pain we feel, the taste of coffee, the anxiety of waiting.

Thomas Nagel articulated this difficulty with great clarity in 1974, in his famous essay addressing the question: What is it like to be a bat? We can know a great deal about bats' echolocation, their neurons, and the sound frequencies they emit. But knowing all this is not the same as knowing from the inside what it means to experience the world as a bat. Consciousness, Nagel writes, is what makes the relationship between mind and body truly intractable. Not because science is useless, but because an external description and a firsthand experience are simply not the same thing.

Schrödinger would likely have recognized this difficulty as an important sign. Not a call to abandon scientific research, but a limit imposed on any theory that claims to have exhausted reality simply because it has explained its mechanisms. We can explain how a nerve impulse travels along a fiber, how a synapse modifies its effectiveness, how an anesthetic alters brain activity. All of this is essential. But it still does not answer—at least not directly—the question: why are these processes accompanied by an experience?

This is where the risk of reductionism lies. Reduction is not always a mistake: it is one of science's most powerful tools. To understand a cell, we isolate a protein; to understand an electrical phenomenon, we distinguish between charges and fields; to understand a disease, we identify a mechanism. The problem arises when an effective method is mistaken for a comprehensive philosophy. If something can be analyzed into its constituent elements, then—it is concluded—it is nothing more than the sum of those elements.

But a symphony is not nullified by the analysis of air vibrations; a poem is not refuted by the chemistry of ink; a beloved face is not exhausted by the anatomical description of its features. The analyses are true, but they are not the whole story. Explaining the material conditions of an experience does not mean we have already explained what that experience means to the person living it.

Eugene Wigner, a 1963 Nobel laureate, expressed this concern in a more daring way than is generally accepted by most physicists today. In a 1961 essay, he argued that it was not possible to formulate the laws of quantum mechanics in a fully consistent manner without reference to consciousness. Wigner went so far as to hypothesize that consciousness played a role in the transition from quantum possibilities to a definite outcome.

This proposal is well-known, but it should not be presented as an established fact. Today, most physicists do not consider it necessary to introduce consciousness to understand quantum measurement. There are different approaches: some focus on interaction with the environment, others envision many possible branches of events, and still others treat the wave function as a tool for organizing probabilities. None of these approaches has put an end to all debate, but none of them justifies claiming that a human thought causes the Moon to appear in the sky.

Wigner's contribution, for our purposes, lies not so much in his solution as in his courage not to treat consciousness as a negligible detail. Schrödinger was even more radical on this point. He did not want to attribute to individual consciousness any special power over particles. Rather, he sought to challenge the very notion of an individual consciousness enclosed within a skull and set in opposition to an entirely external world. Perhaps its strength lies not in providing us with an answer, but in changing the way we look at the question. If consciousness were truly only a private property, produced and possessed by isolated individuals, the world would be a crowd of incommunicable inner worlds. If, on the other hand, the self were a local manifestation of something vaster, then the distance between beings would not be the ultimate truth, even as it remains concrete and often painful.

We cannot prove this second hypothesis using the language of physics. But we can recognize that it is not a fantasy conceived in opposition to science. It is a question that arose even within one of the most rigorous minds of the twentieth century: that of a man who helped found wave mechanics, who reflected on the genetic code before the discovery of the double helix, and who never stopped wondering who the subject hidden behind every equation might be.

The observer, then, is not an accident placed at the end of the cognitive process. It is not an embarrassing presence that science must eliminate in order to become objective. It is the point from which every known world takes sh . It remains to be seen whether this point is truly multifaceted to its very core, or whether its multiplicity resembles the ripples on a single surface of water: real, distinguishable, often far apart from one another, but never completely separated from the water that makes them possible.

One consciousness, many perspectives.

Each of us has a very precise experience of our own separateness. I inhabit this body, not yours; I remember my childhood, not yours; I feel the pain in my knee, not that of my neighbor. My consciousness seems to have boundaries as clear-cut as my skin. It is here, within this particular story, with this name, this voice, these fears, and these expectations.

There is nothing naive about this reality. On the contrary, it is the starting point from which we normally construct the idea of a person. Laws, medicine, education, and love itself all presuppose distinct individuals. Every human being is vulnerable in a unique way; every life story contains wounds that cannot be transferred to another; every responsibility requires someone who can say, unequivocally, "I did it."

Yet Schrödinger believed that this obvious fact, while real in terms of everyday experience, was not necessarily the final word. The plurality of consciousnesses, in his view, might be the way in which a single conscious reality appears through countless organisms, times, and perspectives.

It is a dizzying assertion, and it is best to clear up a few misunderstandings right away. Schrödinger is not arguing that people are illusions in the trivial sense of the term. He is not saying that individual suffering is irrelevant because, after all, "we are all one." Nor is he proposing some sort of collective mind, akin to an invisible network in which thoughts pass directly from one brain to another. His is a more ancient and more radical question: what would happen if the differences between individuals were real as differences in perspective, but not definitive as absolute separations?

In 1956, during the *"Tarner* Lectures" held at Trinity College, Cambridge, Schrödinger returned to questioning the relationship between matter and consciousness. By then, he was a renowned physicist, a Nobel laureate since 1933, and one of the architects of quantum mechanics. But his research had never been limited to physics in the narrowest sense. He was troubled by the idea that science could describe the universe with great precision, yet leave out the very presence of those who experience the universe.

In those lectures, Schrödinger formulated a thesis that would continue to raise questions: consciousness, he wrote in essence, is a singular entity whose plural form we do not truly know. We speak of "many consciousnesses" because we encounter many living beings, many bodies, many voices, and many behaviors. But conscious experience, viewed from within, never presents itself as a crowd. It always presents itself as a unified center of presence.

Of course, every human being can say the same about themselves. And this is precisely where the problem arises. If I can directly know only my own experience, while yours appears to me through words, gestures, and actions, how do I move from the certainty of a single lived consciousness to the certainty of many separate consciousnesses? Normally, we make this transition without hesitation, and for good practical reasons: we recognize an inner life in others because we see that they suffer, desire, remember, and respond. But Schrödinger suspected that the distinction between "my consciousness" and "your consciousness" belonged primarily to the realm of the individual forms through which consciousness manifests itself.

Experimental physics has no say here. No quantum experiment proves that a single universal consciousness exists. The equations of qu l mechanics contain no proof of the spiritual unity of humankind, nor do they authorize the claim that the mind creates the world through an act of will. This clarification is decisive. Physics shows us that, at the microscopic level, the observer, the measuring apparatus, and the phenomenon are not always separable in the simple way imagined by classical physics. But the leap from this to the thesis of a single consciousness is philosophical, not experimental.

Schrödinger takes that step because he feels that the absolute separation of minds is, in turn, a hypothesis—one of the most spontaneous and powerful hypotheses of our civilization, but still a hypothesis.

We can try to grasp the problem through a work that has nothing to do with physics: "*The Ambassadors*," the famous painting that Hans Holbein the Younger created in 1533, now housed at the *National Gallery* in London. At first glance, the painting depicts two richly dressed men, astronomical instruments, a globe, books, and a lute. It is a scene constructed with the almost intellectual precision of the Renaissance. But at the bottom of the canvas appears an elongated, incomprehensible shape, like a suspended smudge.

Only by moving to the side, viewing the painting from an oblique angle, does that smudge reveal its true form: it is a skull. The skull had not been hidden by the painter. It was always there. And yet it does not appear when viewed head-on; it requires a different vantage point. The canvas has not changed: what has changed is the perspective from which the canvas is viewed.

The image, of course, does not prove Schrödinger's view. A metaphor is not a proof. But it can help us understand that the same reality can present profoundly different aspects depending on the vantage point from which we encounter it. Perhaps—and this is the question—what we call "I" is a particular angle from which experience looks out upon the world. Not a fiction, not a mistake, but not the full measure of who we are either.

A child looking at the sea for the first time, a biologist studying its plankton, a fisherman who knows its currents, a woman who has scattered the ashes of a loved one there: are they perhaps seeing four different seas? In a sense, yes. The sea is not identical in each person's experience. It carries memories, fears, data, smells, and different languages. Yet, it is not necessary to think that four separate seas exist. There is a single expanse of water that lends itself to different relationships.

Schrödinger's proposal is bolder: perhaps consciousness, too, resembles this. Not as an object extending through space, of course, but as a reality that expresses itself through many individual openings. Every living being would be a window, not a definitive prison.

One might rightly object to this image by pointing out that windows do not suffer. Human beings do. And precisely for this reason, we must prevent the idea of unity from becoming a consolatory formula. If a person loses a child, they do not need to be told that the pain is only apparent because everything is one. Such a statement would not be profound—it would be cruel. The plurality of lives has concrete consequences, and no spiritual insight can erase them.

The unity Schrödinger speaks of—if we take it seriously—does not eliminate difference: it makes it even more difficult to treat with indifference. If the other is not merely a stranger, if their life is not completely separate from the fabric that sustains us all, then their suffering cannot be dismissed as a distant occurrence.

The British writer and philosopher Iris Murdoch, in her 1970 work *_The Sovereignty of_ Good*, defined love as _"the exceedingly difficult awareness that something other th r ourselves is real."_ This statement does not belong to the tradition of "Advaita Vedānta" nor does it express Schrödinger's thesis. It says almost the opposite, at least on the surface: it reminds us of the irreducible reality of the other. But perhaps this is precisely where the most delicate point emerges. A true intuition of unity should not cause the other to vanish within us; it should make us better able to recognize them.

Unity without respect for difference becomes a refined form of selfishness. It is the risk faced by those who say "we are all one" but no longer listen to anyone. Plurality without any sense of connection, on the other hand, can transform the world into a sum of competing solitudes. Schrödinger seeks a path that erases neither one nor the other: many real lives, many unique perspectives, and yet a source deeper than separation.

Literature has often intuited this tension with greater freedom than philosophy. Fernando Pessoa, who died in Lisbon in 1935, did not merely sign his works with a pseudonym. He invented true "heteronyms": imaginary authors endowed with their own style, biography, character, and worldview. Alberto Caeiro, Ricardo Reis, and Álvaro de Campos were not mere literary masks; they were distinct voices, almost inner selves.

A poem attributed to Álvaro de Campos contains a famous line: "To _be plural like the universe!_" Pessoa was not developing a theory of a single consciousness. He was a poet and knew from the inside the multiplicity of the modern self: the human being is never entirely one, unified, or transparent to oneself. Within us coexist different ages, incompatible desires, memories we no longer wish to acknowledge, and possibilities that never came to life.

Psychology and neuroscience, for their part, have provided many reasons to be wary of the image of a monolithic self. Studies of patients who underwent surgery to sever certain connections between the two cerebral hemispheres to treat severe forms of epilepsy revealed, especially beginning in the 1960s, surprising behaviors. Roger Sperry and Michael Gazzaniga observed that, under certain experimental conditions, the two hemispheres could process information partially independently.

This does not mean that every person has “two consciousnesses” in the simple and sensational sense sometimes found in popular headlines. Those studies are complex and concern specific clinical situations. They do, however, reveal something important: even the psychological unity we attribute to ourselves is a delicate construct, continually shaped by the integration of memory, language, perception, the body, and our relationships with others.

Science, therefore, does not confirm Schrödinger’s thesis of a single consciousness. If anything, it shows us just how difficult it is to define precisely where the subject begins and where it ends. The brain is indispensable to mental life, based on everything we know today; but the fact that there is a very close correlation between the brain and experience does not, on its own, resolve the question of why experience exists. We can describe neural circuits, record electrical activity, and identify areas involved in attention or memory. Yet the fundamental and unfathomable fact remains that red is not merely a wavelength, pain is not merely a nerve signal, and a melody is not merely a sequence of vibrations.

From this unbridged gap arises Schrödinger’s hypothesis. Not as a mandatory answer, but as a possibility for thought. Consciousness might not be produced, out of nothing, by every single organism; rather, it might manifest itself in each organism under unique conditions, like the same piece of music performed on different instruments. The violin is not the piano, and the piano is not the flute. The timbre, the duration, even the breath of the performance change. But what passes through them cannot be reduced to the instrument.

This, too, remains an image, not a scientific theory. And we must preserve its nature as an image, without forcing it to become a false certainty. Its power does not lie in proving that separation does not exist. It lies in making our conviction that we are absolute islands less obvious. Perhaps we are indeed distinct individuals, destined never to fully cross the threshold of another's experience. Or perhaps individuality is a necessary, yet temporary, form through which a broader reality becomes self-aware in multiple ways. We cannot settle the question with an experiment or a formula.

We can, however, realize that the way we respond changes our perspective. If the self is merely a fortress, the other will always be a stranger from whom to defend ourselves or whom we can use. If the self is a window, the difference does not disappear, but the world begins to seem less alien. And perhaps the question Schrödinger leaves us is not: "How can I prove that we are one?" It is more sober and more demanding: "What happens to my life if I stop, at least for a moment, thinking that separation is the only possible truth?"

The Risk of Reductionism—

Reductionism stems from a virtue of science, not from a flaw. To understand a complex phenomenon, science breaks it down. If we want to understand how a heart works, we study the muscle cells, the electrical impulses that regulate its rhythm, the valves, the blood, and the molecules involved in contraction. If we want to understand a disease, we look for specific causes: a bacterium, a mutation, a protein that no longer performs its function.

This method has changed human history. It would be foolish to deny it. Modern medicine, electronics, communications, and much of our ability to alleviate pain and predict natural phenomena depend on the patience with which we have learned to examine the parts.

The problem begins when a method of inquiry becomes an absolute statement about reality. When we move from the entirely reasonable assertion that "a conscious experience depends on the brain" to another, far more sweeping one: "a conscious experience is nothing other than brain activity." There is a decisive difference between these two statements.

The first describes a correlation that neuroscience confirms in countless ways. A head injury, an anesthetic, a lesion, a psychotropic substance, or electrical stimulation can alter memory, mood, perception, body awareness, and even personality. This is not a marginal discovery: it shows just how much the our inner life is tied to the living matter of the brain. The second statement, however, is not an experimental finding in the same sense. It is a philosophical interpretation. It asserts that, once nerve cells and their interactions have been described, there is nothing left to understand. Pain would be merely a burst of signals; music, merely a succession of frequencies; love, merely a set of chemical reactions selected by evolution.

The decisive word in these statements is almost always the smallest one: "merely." Francis Crick, one of the discoverers of the double-helix structure of DNA, published a book in 1994 with a deliberately provocative title, **The Astonishing Hypothesis**. In it, he argued—in a phrase that has since become famous—that our joys and sorrows, memories, ambitions, sense of personal identity, and even the idea of free will are *"nothing more than the behavior of a vast collection of nerve cells and their associated molecules*."

Crick was neither a superficial thinker nor an enemy of humanity. He was a great biologist, convinced that research on the brain would dispel many age-old mysteries. And he was certainly right on one fundamental point: we cannot speak seriously about the mind while ignoring the brain. Any romantic image of a consciousness suspended above biology, completely immune to the body, conflicts with what we know.

But we might ask whether Crick's formula skips over a step too quickly. To say that a symphony depends on air vibrations is not the same as saying that a symphony is fully explained by those vibrations. To say that a novel exists thanks to ink, paper, or the pixels on a screen is not the same as saying that *The Betrothed* is reduced to the chemistry of ink or the structure of a digital file. Material conditions are necessary. But necessary does not mean sufficient.

In 1931, the philosopher and linguist Alfred Korzybski coined a phrase destined to enter the lexicon of contemporary culture: *"The map is not the territory."* A map can be accurate, useful, even indispensable. It indicates boundaries, distances, elevation changes, roads, and rivers. But it does not contain the scent of a pine forest after the rain, the exertion of a trail, the shifting wind on the ridge, or the fear or joy of those who travel it. There is nothing wrong with the map. Its limitation begins only when we mistake it for the entire landscape.

Even the scientific description of the brain is a map of extraordinary precision, continually corrected and refined. It shows us neural networks, chemical exchanges, and regions involved in attention, memory, and the regulation of emotions. But the question Schrödinger never stops asking concerns precisely what the map, by its very nature, risks leaving in the background: why are these processes accompanied by a lived experience?

We are not simply asking how the brain distinguishes red from green. This problem can be addressed by studying the retina, the visual pathways, and the brain areas that process color. We are also asking what the "redness" of red is when it appears to consciousness: that immediate quality that no data table seems to contain.

This is not an objection to science. It is the recognition that science, to achieve its full potential, adopts a particular perspective. It measures what can be shared, repeated, and observed from the outside. And it is right to do so. But consciousness has a side that cannot be entirely entrusted to an external perspective: the fact that it is experienced by someone.

A doctor can rigorously describe a migraine attack: the progression of symptoms, the inflammatory and h processes, the activation of certain neural pathways, and the effectiveness of a medication. Yet even the most comprehensive description does not convey the experience of pain to the reader. Not because medicine has failed, but because knowing the mechanism of something and being inside that something are two different forms of knowledge.

Michael Polanyi, a Hungarian-born British chemist who later became a philosopher of science, expressed this difficulty in a simple phrase: *"We know more than we can say.*" He wrote this in 1966, in **The Tacit* Dimension*, a book dedicated to that part of knowledge we possess without being able to fully translate it into explicit rules.

Polanyi was thinking, for example, of the face of a loved one. We can recognize it instantly among hundreds of other faces, but we often cannot say exactly which specific details allow us to do so. We can list the color of the eyes, the shape of the mouth, the hairline; and yet the identity of that face seems to elude the sum of those details. It is not magic. It is a form of concrete, embodied understanding that precedes our explanations.

The same thing happens when a skilled pianist performs a piece by Chopin. The notes are written down, the tempo is indicated, and the dynamics can be carefully notated. But no score, however precise, guarantees the interpretation. The sound arises from hands, breath, memory, technique, listening, hesitation, and presence. Reducing the performance to the laws of acoustics would be correct, but incomplete. The laws of acoustics tell us how a sound propagates; they do not tell us what happens when that sound becomes nostalgia, anticipation, consolation, or unease in the listener.

Here we arrive at the point that interests Schrödinger. Physics is not false simply because it does not address inner, , experience. It would be unfair to demand of it what it has not set out to do. Modern physics has built its power precisely by methodically setting aside individual experience: color becomes frequency, heat becomes the motion of particles, sound becomes vibration, and the body becomes mass and energy.

This choice has been fruitful. But Schrödinger observed that the world thus obtained, though extraordinarily effective, risks appearing devoid of the one who knows it. We have a description of the cosmos in which the subject who looks, feels, fears, and hopes seems to find no place except as a secondary effect of impersonal processes.

The paradox does not lie in the fact that science is incomplete—every science is, and that is precisely why it continues to seek. It lies in the fact that we, as conscious beings, produce a description of the world that tends to render consciousness itself invisible. Schrödinger did not offer a definitive solution to the mystery of life; he posed questions capable of bringing the mystery into sharper focus. How can living matter preserve and transmit such a complex order? How does an organism resist, for a certain time, the disorder that seems to dominate everything? His deepest lesson, perhaps, did not lie in the individual answers, some of which would be superseded by subsequent developments in biology. It lay in the refusal to believe that an explanation, in order to be rigorous, must necessarily impoverish what it explains. It is possible to understand the molecular basis of heredity without thinking that a person is merely their genome. After the completion of the first draft of *the "Human Genome Project,"* announced in June 2000 and published in broad outline in 2001, many expected a sort of definitive dictionary of the human being. But the reality turned out to be more complex. Knowing which genes we possess is extremely important for understanding predispositions and diseases; yet it does not provide us with the story of a life, the consequences of an encounter, the weight of a choice, or the uniqueness of a loss or a commitment. A gene does not write a poem. Not because biology is irrelevant to poetry—a brain, a body, a learned language, an evolutionary history are all necessary conditions—but because the meaning of a verse cannot be found by isolating a molecule. Reduction is a valuable tool when it helps us see the components. It becomes reductionism when it claims that the components are all there is to understand. There is a similar difference between knowing the chemical composition of a fresco and looking at Giotto's frescoes in the Scrovegni Chapel in Padua. A restorer must know which pigments were used, how the color adheres to the plaster, and what deterioration processes threaten the surface. Without this knowledge, the work could not be preserved. But no analysis of the pigments can fully capture the experience of entering the chapel and having one's gaze meet the deep blue of the vault or the faces of Anna and Joachim as they embrace near the Golden Gate. Scientific analyses do not diminish the value of the painting. They simply do not coincide with the experience of the painting.

The cultural risk of reductionism is therefore not precision, but the loss of layers. A biological explanation of a fear may be true; a psychological explanation may be true; an interpretation linked to a person's biography may be true; sometimes even a question about the meaning that fear takes on in that person's life may be necessary. These perspectives need not compete with one another to establish which is "the only real one." They can shed light on different aspects of the same event.

Of course, not all interpretations are equal. It is not enough to invoke mystery to evade evidence, nor to use the word "consciousness" as a free pass for any belief. Pseudo-treatments, claims of spiritual healing that encourage people to abandon effective treatments, and fantasies that quantum mechanics supposedly proves every form of magical thinking—none of this is openness. It is confusion. Scientific prudence is a form of respect for reality. But even respect for reality requires that we not believe that only what we can measure from the outside is real.

When a mother watches over her feverish child through the night, specific processes take place in her brain. Certain hormone levels rise or fall; circuits related to attention and alertness are activated; sleep deprivation alters cognitive abilities. All of this is true and can be of great importance. But there's more to the scene than that. There is a person waiting for morning, listening to another person's breath, feeling fear and offering care. There is a meaning that does not float outside of matter, yet is not entirely captured by the language of molecules either. Perhaps the point is not to decide whether a human being is matter or spirit, brain or consciousness. Such clear-cut oppositions may reassure us, but they often oversimplify things. The point is to ask ourselves whether matter, when it becomes living and conscious, does not reveal possibilities that even our most basic language still struggles to accommodate.

Consciousness Without Boundaries? From Schrödinger to Contemporary Theories.

Among Erwin Schrödinger's deepest insights is one that surpasses even the fame of his cat, quantum superposition, and the measurement problem. It is an insight that concerns us directly: consciousness is not a multitude of isolated realities, but a single presence that manifests itself through countless living forms.

We are accustomed to imagining ourselves as separate minds. Each of us holds our own inner world—a story, memories, wounds, desires. My pain is not your pain; my thoughts are not accessible to your immediate experience. This separation is real at the level of individual life. But it is not necessarily the ultimate truth about who we are. Schrödinger expressed this with a famous formula: *"Consciousness is a singular whose plural is unknown."* In the deepest sense, there are not many consciousnesses, each enclosed within its own body. There is a single consciousness—an original reality of experience—that views the world from billions of perspectives.

This is neither a poetic concession nor a vague form of spirituality applied to science. It is a precise philosophical thesis: what we call "I" does not produce consciousness, but constitutes a local configuration of it. The individual is real, but is no more separate from the source of experience than a wave is separate from the ocean.

The brain, the body, memory, and personal history are not illusions. They are the concrete ways through which consciousness takes on a specific form. Each of us is a unique window: a biological, emotional, and cultural- al structure through which conscious reality coalesces into an individual perspective. There are many windows; there is one light.

This image does not reduce the human being to a mere passive antenna. The brain is not an insignificant object, nor is it an accessory to the mind. It is an extraordinary organism of selection, organization, and manifestation. It makes it possible for universal consciousness to become this face, this voice, this memory, this person. But this does not mean it should be considered the absolute creator of consciousness, as if experience could emerge from nothing through processes completely devoid of interiority.

Here lies one of the central cruxes of the problem of consciousness. Neuroscience describes with increasing precision the correlations between brain activity and experience: we know that certain lesions alter language, that anesthesia interrupts waking consciousness, and that certain neural networks are involved in memory, attention, and perception. All of this is indisputable and fundamental. But a correlation, however close, is not equivalent to a ultimate explanation.

Knowing which brain circuits are activated when we see the color blue does not explain why blue is experienced. It does not explain why an electrochemical process is accompanied by the luminous quality of a color, the taste of coffee, the pain of a loss, or the silent presence of being here. Science can describe the mechanisms of experience; the decisive question is what makes experience itself possible.

David Chalmers has called this enigma *the "hard* problem" of consciousness. It is not difficult because of a lack of data, but because no amount of objective data seems to automatically translate into subjectivity. We can map every synapse in a brain and measure every electrical current , but one question remains: why is all of this experienced from within?

The field theory of consciousness begins here. Consciousness is not a late and accidental product of organized matter; it is a fundamental dimension of reality. Like space, time, energy, or information, it belongs to the deep structure of the universe. Matter does not generate consciousness out of nothing: it channels it, modulates it, integrates it, and gives it an individual form.

From this perspective, the cosmos is not an immense, blind mechanism that, after billions of years, happens to produce a few sentient beings by chance. Rather, it is a reality in which interiority and exteriority have been present from the very beginning. What physics observes from the outside—matter, energy, fields, and relationships—also possesses an inner dimension: the possibility of experience.

The field of consciousness should not be trivially imagined as an invisible cloud suspended in space, analogous to a magnetic field or a form of physical radiation that has not yet been measured. It is more fundamental than that. It is not an object among objects: it is the very principle by which something can appear, be perceived, and become experience. Everything that physically exists manifests itself within this primordial horizon.

Schrödinger arrived at this vision in part through his engagement with Indian thought. In those sources, he found a conception of being that does not rigidly separate the subject from the world or the individual from the whole. The phrase *"Tat tvam* asi" expresses precisely this insight: the deepest self does not coincide with the narrow biographical identity, but belongs to universal reality.

This does not mean that every person loses their uniqueness in an indistinct mass. On the contrary: the one consciousness expresses itself as plurality without becoming absolute division. Every living being is a unique modulation of the field, a point at which the universe perceives itself from a one-of-a-kind perspective.

A melody helps us understand this idea. The notes are distinct; they have different pitches, durations, and timbres. No two notes are identical. Yet, when we listen to them, we do not perceive a random collection of sounds: we perceive a musical unity. Individualities do not disappear into the whole, but acquire their full meaning only within it. Thus, individual consciousnesses are as real as the notes are real; but what makes them possible is the single field of consciousness.

Contemporary theories of the mind, while not always using this language, are moving ever closer to a non-reductionist view. Giulio Tononi's theory of integrated information, for example, starts from the decisive fact that every experience is a unified whole. When we listen to music, observe a room, or remember a loved one, we do not receive disconnected fragments. We experience a coherent, integrated field in which sensations, images, emotions, and thoughts hold together.

Tononi has sought to formally express this unity through the concept of integrated information, denoted by the symbol Phi. A system is conscious to the extent that it constitutes an irreducible unity: something more than the mere sum of its parts. This insight is crucial because it recognizes that consciousness cannot be explained by reducing it to a list of neurons or a series of electrical impulses. Experience is always a whole.

The field theory of consciousness takes this a step further. If an integrated system can harbor a center of experience, it is because experience is not foreign to nature. Integration does not create consciousness out of nothing; it creates the conditions for its organized manifestation. The human brain' , with its immense complexity, does not manufacture the light of consciousness; it constructs a lens capable of focusing it into an extremely rich, reflective, and personal form.

Another proposal, closer to physics in the strict sense but much less accepted by the scientific community, comes from Johnjoe McFadden, a British molecular biologist. Beginning in the 1990s, McFadden hypothesized that the electromagnetic fields produced by neuronal activity might play a causal role in consciousness. According to this idea, the brain is not merely a network of cells that transmit signals; it also produces an electromagnetic field capable of integrating information distributed across many brain areas.

The hypothesis is known as the CEMI theory, short for "Conscious Electromagnetic Information." It has the merit of taking a real fact seriously: all neuronal activity is accompanied by electrical and magnetic phenomena, which we can record, for example, using an electroencephalogram or magnetoencephalography. But the next step—arguing that this field is the seat or decisive vehicle of consciousness—remains controversial. Many neuroscientists believe there is insufficient evidence and that the observed electromagnetic field is an effect of brain activity, not its hidden driving force.

The difference is crucial. An electromagnetic field of the brain is a measurable physical reality; a universal field of consciousness is, for now, a philosophical or metaphysical possibility, not an entity detected in the laboratory. The two ideas may evoke one another, but they should not be conflated.

Even contemporary panpsychist perspectives, advocated in various forms by philosophers such as Galen Strawson and Philip Goff, converge on this essential : consciousness must be part of nature, not appear as an inexplicable miracle in a universe that, up to that point, would have been entirely devoid of inner life.

To say that consciousness is fundamental does not mean attributing human thoughts to stones, chairs, or elementary particles. It means abandoning the naive idea that there exists, on the one hand, completely lifeless and mute matter and, on the other, minds that suddenly come to life. A continuum extends between these extremes. Consciousness takes on different degrees, forms, and levels of integration. In human beings, it reaches a level of complexity capable of language, autobiographical memory, imagination, and self-awareness; but its origin does not begin with us.

This view also changes the way we interpret the relationship between the brain and the mind. Neurological cases clearly demonstrate how much individual identity depends on the integrity of the nervous system. A brain injury can alter personality, impair language, change perception, and transform memory. Anesthesia can temporarily interrupt ordinary conscious experience. Dementias can progressively erode access to one's personal history.

These facts do not refute the realm of consciousness; they clarify how it functions. If a radio is damaged, the music being broadcast does not disappear from the universe; it is distorted, interrupted, or rendered unrecognizable in the device receiving it. Similarly, when the brain is compromised, what is altered is the individual form through which consciousness expresses itself.

The radio metaphor should not be understood as a convenient escape from neuroscientific data. The brain is infinitely more complex than a receiver. It is at once a filter, a translator, an organizer, embodied memory, and a center of action and relationship. But the principle remains: the dependence of individual expression on the brain does not prove that the brain is the absolute origin of consciousness.

The materialist model tends to view the brain as a factory of experience. The field-of-consciousness model, on the other hand, views it as an interface: a place where the universal field becomes an individual subject. This second interpretation has a decisive philosophical advantage: it does not ask matter to accomplish the impossible—namely, to produce subjective experience from a reality defined from the outset as completely devoid of experience.

There are also attempts to identify the brain's electromagnetic field as a possible physical basis for the conscious unit. Johnjoe McFadden has proposed that the electromagnetic fields generated by neurons may play a causal role in experience. This idea does not coincide with the theory of the universal field of consciousness, but it points in an important direction: the mind does not seem reducible to the isolated behavior of billions of cells. Rather, it appears as a global, unified, dynamic property, capable of bringing together processes distributed throughout the brain into a single experience.

The field of consciousness extends this insight beyond the brain. If mental activity possesses a field-like character even within the organism, it is not arbitrary to recognize that conscious individuality is embedded in a broader, continuous, and interconnected reality. We are not sealed monads. We are nodes of experience in a shared fabric.

This assertion has profound ethical consequences. If others are not merely biological objects distinct from us, but expressions of the same conscious reality, then compassion is not merely a moral convention. It is a form of lucidity. Another's pain is not my pain in the psychological and personal sense; I cannot take the place of their experience. But neither is it something entirely foreign. It is the very same light of consciousness that, in another window, takes the form of suffering.

The separation between individuals remains necessary in practical life. Each person is responsible for their own actions; each possesses a vulnerable body, a history, and a name. Recognizing the unity of consciousness does not erase differences, nor does it authorize us to ignore personal boundaries. On the contrary, it makes those boundaries more sacred: every being is a place in which the universe manifests itself from within.

Schrödinger understood that the "I" is not the absolute ruler of reality, but a temporary figure of consciousness. We say "I" and believe we are pointing to an isolated entity; in truth, we are pointing to an interweaving of memory, body, language, relationships, and attention, illuminated by a more primordial presence. That presence does not belong to us as a possession. It is we who appear within it.

A single light, then, many windows. Many faces, many biographies, many voices; but a single field of conscious being that, through each of us, encounters the world and recognizes itself.

VIII. The cat's final lesson: sobriety in the face of reality.

"We can know more than we can say."
Michael Polanyi, *The Tacit Dimension*, 1966.

"Science cannot tell us a word about red and blue."
Erwin Schrödinger, *Nature and the Greeks and Science and Humanism*, 1954.

This chapter does not propose that science surrender in the face of mystery, nor does it invite us to replace rigor with vague spirituality. On the contrary: it attempts to follow through to its logical conclusion the demand for precision that drove Schrödinger and many of the leading physicists of the twentieth century. It is precisely that precision, when it encounters questions about consciousness, lived experience, and the meaning of existence, that teaches us to distinguish what we know from what we imagine we know.

Science describes the regularities of the world with extraordinary power. But it does not possess the world. Its formulas do not exhaust reality, just as a map, however accurate, does not contain the wind in a valley, the smell of the earth after rain, or the fatigue of those who traverse it. This awareness does not give rise to contempt for reason; rather, it gives rise to a more mature reason, capable of questioning without presuming to reduce everything to its own measure.

At the end of a journey dedicated to Erwin Schrödinger, the cat paradox, life, consciousness, and the traditions of India, an understandable temptation may arise: to seek a conclusion that puts every element in its place. Ultimately, we would like to know whether quantum physics truly demonstrates the unity of reality; whether consciousness is the foundation of the universe; and whether the insights of the Upanishads have found confirmation in the language of modern science.

But perhaps this expectation of a final verdict is precisely what Schrödinger's journey invites us to reconsider.

Not because the questions are wrong. They are profound questions, and sometimes unavoidable. The problem begins when we ask a physical theory to provide us—

beyond what it can actually tell us—with consolation, a complete metaphysics, or a guarantee regarding the meaning of existence. Ph l physics is no less profound because it cannot offer us all of this. It is profound, if anything, because it forces us to take the limits of its language seriously.

Schrödinger was not a thinker willing to accept these limits with indifference. He was one of the founders of quantum mechanics and, at the same time, a restless scholar of its meanings. It was not enough for him that the equations worked: he wanted to understand what, in their language, we were actually saying about the world. This restlessness did not lead him to reject science. Rather, it led him to a form of sobriety: the awareness that an exact description does not necessarily coincide with a complete experience.

In this sobriety, we may perhaps find the most meaningful point of contact between science and wisdom. Not an artificial fusion. Not the idea that a laboratory is a temple in disguise, nor that a wisdom text contains particle physics in advance. More modestly—and perhaps more profoundly—both can teach a discipline of attention: to see more clearly, to speak with greater caution, not to confuse the name of a thing with the thing itself.

In 1954, in a short book based on lectures given at University College London, Schrödinger made an observation destined to be memorable: *"Science cannot tell us a word about red and blue."* The statement may seem provocative. Of course, physics can say a great deal about color: it can describe the wavelengths of light, the behavior of photons, the functioning of the retina, the electrical signals transmitted along the optic nerve, and the brain regions involved in perception. Yet all of this does not coincide with the red we experience.

It does not correspond to the intense red of a piece of fabric seen at a market in Jaipur, to the deep red of a sunset on a wint , or to the red of a traffic light that stops us when we're distracted. The physical description offers us measurable relationships and reliable predictions. But the experience of color—how it appears to someone—belongs to another aspect of the question.

Here, physics has its say: visible light occupies a specific region of the electromagnetic spectrum; color perception depends on the interaction between radiation, the eyes, the nervous system, and the brain.

Here the philosophical interpretation comes in: can a third-person description, however detailed, fully capture the first-person experience? Or is there always a gap between the objective account of processes and the way those processes are experienced?

And here, finally, we can turn to an existential question: what does it mean to inhabit a world that we can describe rigorously, without thereby possessing its meaning from the outside?

Schrödinger did not mean to say that color is unreal, nor that it is a private fantasy devoid of any connection to the physical world. His point was more subtle. The scientific method, in order to obtain knowledge that is shareable and verifiable, selects certain aspects of experience: quantities, relationships, measurements, and regularities. This selection is its strength. But, inevitably, it leaves out the immediate quality of sensation.

It is a methodological concession, not a defeat. To measure with precision, science must construct a common language, freed as much as possible from individual impressions. When a physicist speaks of the frequency of radiation, he does not ask what emotion that light evokes. When a biologist studies a cell, he does not ask what it means for us to feel alive. Not because these questions are unworthy, but because they require different tools.

Confusion arises when we transform this necessary delimitation into a comprehensive theory. If science does not record the value of a promise, the beauty of music, or the pain of grief, it does not follow that promise, beauty, and pain are illusions. It merely follows that they are not measurable objects in the same sense that the mass of an electron or the temperature of a gas are.

Werner Heisenberg, who shared with Schrödinger the feverish years of the birth of quantum physics, expressed this with great clarity: *"What we observe is not nature in itself, but nature as it is revealed by our way of questioning it.*" This statement does not justify a relativism according to which everyone invents their own universe. Experiments defy our desires; measurements can prove us wrong; nature is not a blank slate onto which we project any narrative we please.

But the way we ask questions matters. Every instrument makes certain aspects visible and leaves others in the shadows. An electron microscope, a particle accelerator, a space telescope: each vastly expands our view, but does so in a specific way. They are not absolute windows open to everything; they are instruments of inquiry.

This is also true of words. We say "particle," "wave," "cause," "observer," "reality." Ancient words, often born of ordinary experience, that physics is compelled to use to speak of phenomena that elude everyday intuition. The risk is mistaking these words for accurate images. Quantum mechanics does not allow us to imagine atoms as tiny balls endowed with a secret and perfectly defined history. Nor does it invite us to think that the human mind creates whatever it observes at will. Rather, it provides us with a formalism of exceptional effectiveness, whose translation into familiar images remains problematic.

This is the point at which language does not fail: it becomes self-aware. It recognizes that it is indispensable and, at the same time, insufficient.

In 1921, Ludwig Wittgenstein concluded the *Tractatus Logico-Philosophicus* with a famous proposition: *"Whereof one cannot speak, thereof one must be silent."* This phrase is often used as an invitation to stop thinking. But Wittgenstein's silence is not obtuseness; it is the refusal to produce discourses that feign a clarity they do not possess. Not everything that matters can be transformed without residue into a well-formed proposition. Science does not demand silence because it has exhausted its questions. It demands silence, at times, to make room for better-formulated questions.

There is a form of knowledge that resembles conquest. A phenomenon is isolated, measured, reproduced, and predicted. It is the approach that has made much of modern science possible, and it would be naive to view it with suspicion. Thanks to this knowledge, we have understood biological processes, cured diseases, sent instruments beyond the solar system, and built technologies that radically transform daily life.

However, when the act of knowing is applied uncritically to every sphere of existence, it can turn into possession. Knowing a person becomes cataloging their traits. Knowing a forest becomes calculating its resources. Knowing the brain becomes believing we have solved the mystery of consciousness. Knowing life becomes breaking it down into its molecules, forgetting that those molecules are also the body of a living being that is born, suffers, finds its way, and dies.

Schrödinger, in his scientific work and in his reflections on life, never ceased to seek explanations. But he did not confuse the success of an explanation with the right to consider reality definitively captured. His interest in biology did not stem from a desire to reduce life to an insignificant mechanism. It stemmed from a sense of wonder at the improbable order of an organism, at its ability to sustain and renew itself over time.

An organism is not merely matter arranged in a particular form. From a biological perspective, it is matter traversed by processes of organization, exchange, regulation, and repair. From the perspective of experience, it is also a vulnerable center of needs, perceptions, and relationships. These two descriptions need not be at odds with one another. However, it would be reductive to claim that the first cancels out the second.

The philosopher and scientist Michael Polanyi, born in Budapest in 1891 and later a professor at Manchester, articulated this insight in a few words: *"We know more than we can say."* Published in 1966 in **The Tacit* Dimension*, the phrase refers to the tacit knowledge that accompanies many of our actions. We recognize a familiar face before we can list all the details that make it recognizable. An experienced physician sometimes picks up on a clinical sign before being able to translate it into a set of rules. A musician senses that a performed phrase is off-balance, even if they cannot yet explain technically why.

This does not mean that tacit knowledge is infallible. It can deceive us, and that is precisely why science requires public scrutiny, replicable experiments, and the testing of hypotheses. But it does mean that human knowledge is never merely an archive of explicit information. It is also trained attention, bodily memory, familiarity, and the ability to recognize patterns.

Even in the laboratory, knowledge retains this dimension. An experiment is not carried out by an abstract mind. It requires hands, calibrated instruments, learned gestures, and patience in distinguishing a signal from noise. It requires a community that discusses results, corrects errors, and establishes criteria. The famous scientific objectivity is not the erasure of the human: it is a rigorous human endeavor, designed to limit individual illusions.

To know without possessing, then, means not sacrificing precision, but freeing it from the will to dominate. It means observing a phenomenon without expecting it to conform to our conceptual framework. It means recognizing that maps are necessary, while it would be dangerous to forget that they are merely maps.

This approach bears a striking resemblance to certain wisdom traditions, but it is wise to proceed with caution. The Upanishads and "Advaita Vedānta" are not scientific protocols; they offer no experimental predictions nor do they replace the work of physics. Their language concerns liberation from ignorance, the nature of the self, and the experience of unity. Science, on the other hand, subjects its claims to public criteria of verification and review.

Yet, we can discern a common attitude: a wariness toward the ego that seeks to place itself at the center and turn everything else into an object. In different ways, the honest researcher and the practitioner of a contemplative path can learn the same discipline: not to force reality to confirm what we already wish to believe.

When Schrödinger found a source of inspiration in ancient India, he was not seeking from it a technical solution to the problems of quantum theory. Rather, he was seeking a vocabulary and a perspective capable of challenging the image of the self as a separate, self-sufficient entity, definitively enclosed within its own boundaries. This is a philosophical and spiritual choice, not an inevitable conclusion of the equations. Saying this does not diminish its fascination. It makes it more serious.

Physics can show us that our customary images of matter are insufficient. It can teach us that observation is not a trivial, neutral act. It can reveal a web of relationships where common sense saw isolated objects. But the transition from physical interdependence to the spiritual unity of being remains an interpretive leap. It can be fruitful, it can be debated, it can deeply move someone; yet it must not be disguised as a laboratory result. Wisdom perhaps begins precisely at this point: in not attributing to a discipline what it has not said, and in not depriving other questions of their dignity simply because they do not conform to the procedures of physics.

The word "silence" can evoke many images: a bare room, a monastery, a forest, a library. But the silence we need here is not, first and foremost, a place. It is a quality of attention.

It is the silence that precedes an experiment, when the researcher knows they must let the data correct their hypothesis. It is the silence that follows a great theory, when, instead of turning it into an all-encompassing legend, we ask ourselves where it applies, what it truly explains, and what problems it leaves unresolved. It is the silence in the presence of another person, when we realize that no psychological definition can fully capture a life.

It is not an absence of speech, but a renunciation of superfluous speech. It is not an escape from thought, but thought that has ceased to confuse its own voice with the voice of the world. Simone Weil, a French philosopher who died in 1943 at the age of just thirty-four, called this disposition "attention." "*Attention, taken to its highest degree, is the very same thing as* prayer," she wrote. One need not share her religious language to grasp its power. Authentic attention does not immediately make use of wh t encounters; it does not pigeonhole it to defend itself against its otherness. Instead, it exposes itself to the possibility of being transformed.

In this sense, science, too, is an exercise in attention. Good research does not start from the certainty of already being right. It starts from a hypothesis, of course, but accepts that an unexpected result may require a change of course. The history of science is dotted with corrections of this kind. It is not a linear path toward absolute possession of the truth; it is a practice of continuous refinement.

In 1927, at the fifth Solvay Conference in Brussels, the leading figures of modern physics engaged in a heated debate over the meaning of quantum mechanics. In that famous photograph, we see Einstein, Bohr, Heisenberg, Planck, Curie, de Broglie, Dirac, Pauli, and other key figures of an irreplaceable era. They were not mystics lost in contemplation. They were scientists in disagreement, striving to understand what their theories actually allowed them to say.

The value of that meeting does not lie in the fact that a final word was spoken. It was not. Rather, it lies in the ability to engage with a problem without dismissing it with a rhetorical formula. Einstein was not convinced of the completeness of Bohr's interpretation; Bohr did not abandon his own position. Yet dissent, when it remains faithful to the facts and to an intelligent understanding of others' reasoning, can be a noble form of respect for reality.

Today, in an age that rewards immediate opinions and all-encompassing explanations, this lesson seems particularly valuable. Quantum physics has become a field ripe for grand claims: "consciousness creates reality," "everything is energy," "science has proven what the ancients knew." Statements like these can be seductive because they seem to unite the need for knowledge and the need for meaning into a single image. But their simplicity often hides a cost: the abandonment of distinctions.

Physics does not prove that every desire alters the external world. It does not prove personal immortality. It does not prove that the universe has a recognizable moral purpose. Nor does it prove, strictly speaking, a spiritual doctrine of unity. Those who claim otherwise are not upholding the rigor of science: they are using it as a garnish for a belief that has another origin.

And yet it would be just as impoverished to conclude that, since physics does not prove these meanings, they are therefore without value. The question of meaning cannot be dismissed by a negative experiment. Pain, love, death, beauty, and responsibility do not become any less real simply because they lack a universal unit of measurement.

Perhaps silence is precisely the space in which these questions can remain alive without being filled in too hastily. There is a silence of science: the one that sets in when results do not warrant further conclusions. There is a silence of philosophy: the one that acknowledges the inadequacy of words in the face of certain fundamental experiences. There is a silence of contemplative traditions: not as a rejection of the world, but as a suspension of the self that constantly seeks to comment on, judge, and appropriate whatever it encounters.

These silences are not identical and should not be confused. It would be artificial to declare them to be the same thing. But they can converge in a common gesture: the refusal to pretend that the mystery has already been solved. Schrödinger leaves us, ultimately, with this sense of restraint. His physics contributed decisively to changing the face of science. His reflections on life and consciousness have continued to challenge philosophers and scientists. His dialogue with Indian thought remains complex, personal, and at times debatable, but no less relevant for that. In all of this lies a rare kind of fidelity: the refusal to separate intellectual precision from the question of what it means to be h e here. We do not receive a definitive doctrine from him. We receive something more demanding: an invitation not to use science as a new mythology and not to use mystery as an excuse to abandon clarity.

We can continue to study, measure, discuss, and refine our theories. We can also continue to ask ourselves what consciousness is, what connects a living being to the rest of the cosmos, and what place our freedom has in a universe governed by laws. But perhaps we will have to learn to do so differently: with less haste to reach conclusions, less desire to possess, and more attention to what things reveal and to what, despite everything, they continue to hold in silence.

The map is not the forest, but it can teach us how to walk.

Modern science has achieved its greatest results by learning to isolate, measure, compare, and predict. It is a discipline of precision, and it would be naive to set it against a vague celebration of mystery. Yet precisely those who know scientific work from the inside know that an explanation is not equivalent to a definitive conquest of reality.

At this point in our journey, the question is not whether we should seek knowledge or not. The question is more subtle: what do we do with knowledge? Do we use it to make ourselves less blind, more responsible, more capable of care? Or do we transform it into the claim that we have the world in the palm of our hand? Schrödinger, a rigorous physicist and attentive reader of Indian traditions, guides us toward a possible answer: knowing can also be a way of participating in what we study, without reducing it entirely to an object at our disposal.

There is a form of knowledge that resembles a grasp. We look at something, separate it from its context, measure it, classify it, and identify its laws. Then, often without realizing it, we take a further step: we imagine that what we have described is now entirely ours. As if giving something a name meant having exhausted its reality.

Modern science also arose from this act of separation. To understand the motion of the planets, Galileo had to free himself from many preconceived notions. To study the human body, anatomy had to open up what reverence had rendered untouchable for centuries. To understand diseases, medicine had to observe organs, tissues, microbes, and molecules. There is nothing contemptible about this: on the contrary, a decisive part of our ability to cure, prevent, and predict depends on the patience with which we have learned to break down phenomena.

The problem begins when a method—effective in one field—becomes a totalizing philosophy. When what can be measured becomes the only thing we consider real. When knowledge is confused with possession.

To know a forest, then, risks meaning nothing more than calculating its timber volume. To know an animal, nothing more than recording its behavior and genetic makeup. To know a person, nothing more than listing their age, character, habits, traumas, and hormone levels. All this information may be true and important. But it is not necessarily the person, the animal, or the forest.

It is one thing to have data about something; it is quite another to have truly engaged with what that data seeks to describe. Erwin Schrödinger was certainly no enemy of explanation. His scientific life was dedicated, with extraordinary rigor, to finding mathematical forms capable of making the behavior of matter intelligible. Yet, when he reflected on life in the 1940s, he sensed that living beings demanded a broader perspective. In 1944, in lectures delivered at *Trinity College* in Dublin, Schrödinger addressed the problem of biological order: how can an organism maintain its structure, renew itself, and resist the disorder that seems to threaten everything?

His question was physical, not mystical. But it was not a cold question. Behind it lies the wonder and in the face of the elementary fact that a cell manages to retain a form, that a seed becomes a plant, that a wounded body heals itself, that a living being maintains its fragile coherence for a certain time amid the ceaseless flow of matter.

From this perspective, life is not an exception to the laws of nature. And yet it is not merely a sum of its parts either. An organism is made up of molecules, certainly; but in a living being, those molecules form a web of relationships, exchanges, rhythms, and regulatory processes. To say that a symphony is "merely the vibration of air" is not false; it is simply insufficient. It describes a real level, but leaves out the musical order, the form, the act of listening, and the effect that music produces in those who receive it.

The difficulty does not lie in choosing between explanation and experience. It lies in not letting one override the other. An experienced doctor, upon entering a hospital room, can tell from a patient's complexion, posture, breathing, and the way they answer questions that something is wrong—even before having all the necessary test results. Of course, he'll need to verify his impression: order tests, compare the data, and avoid jumping to conclusions. But that initial impression isn't necessarily a whim. It may be the result of years of trained observation, thousands of encounters, corrected mistakes, and details that his experience has taught him to recognize without immediately translating them into a formula.

The same is true in many professions. A restorer may notice that a painted surface "doesn't respond" as it should; a musician recognizes a slight discord that an inexperienced listener doesn't perceive; a sailor knows how to read a shift in the wind before the instruments clearly indicate it. These are not mysterious powers. It is embodied knowledge: a competence that flows through the body, memory, practice, and familiarity.

Science itself, when truly practiced and not merely described in textbooks, is full of this dimension. An experiment is not carried out by an abstract consciousness suspended above the world. It requires calibrated instruments, skilled hands, patience, and the ability to distinguish a signal from background noise. It also requires a community: colleagues who discuss, replicate, critique, find errors, and modify procedures.

Scientific objectivity does not arise from the elimination of the human element. Rather, it arises from a disciplined use of the human element: from the construction of shared methods that limit our preferences, our illusions, and our tendency to see what we hope to find. This is already a form of humility. The best scientist does not say, "Reality must confirm my idea." Instead, they say, "Let's see if my idea holds up against reality." And if the results contradict it—at least in principle—they are called upon to change their mind.

Knowing without possessing begins here: in recognizing that what we study is not obliged to fit meekly into our frameworks.

An episode from the history of ethology can help us illustrate this difference. In July 1960, a young British researcher named Jane Goodall arrived at the Gombe Stream Reserve, in what is now Tanzania, to observe chimpanzees in the wild. She had no traditional academic training in zoology; she was working under the guidance of anthropologist and paleontologist Louis Leakey. For months, the animals fled at the sight of her. Goodall had to learn to wait, to stay still, and to get to know the place and the animals' rhythms before she could hope to understand them.

Then she observed a chimpanzee she named David Greybeard using a blade of grass to extract termites from their nest. It was not merely a curious detail: the use and rudimentary manufacture of tools—which many at the time considered a uniquely human trait—unfolded before her eyes in a species closely related to our own.

The episode was significant for science. But perhaps it also teaches us something else. Goodall did not arrive at that discovery by subjecting the animals to a laboratory-designed experiment. She arrived at it through a long-term presence, patient attention, and a willingness to learn from the chimpanzees' behavior.

Of course, even that kind of observation has its problems. The observer's proximity can alter the animals' behavior; the practice of assigning individual names, rather than numbers, was criticized by some scientific circles at the time as being too intrusive. But this very debate is instructive. There is no such thing as a completely detached, pure, uninvolved perspective. All knowledge arises from a particular standpoint, from tools, from questions, and from a certain distance or proximity. The key is to learn to acknowledge this, not to pretend it does not exist.

Participating, therefore, does not mean becoming one with what we study. It does not mean giving up distinction, criticism, or verification. It means accepting that knowing is always, to some extent, entering into a relationship. Even the etymology of the word "participation" suggests something similar: to take part. Not to take the whole. Not to dominate it. To take part in a reality that precedes and transcends us.

This idea does not require us to idealize nature. Nature is not always harmonious, nor is it benevolent. It encompasses predation, disease, struggle, and death. A forest is not merely a quantity of absorbed carbon, a surveyed habitat, or a timber reserve. It is also a place of relationships that make life possible for many species— d, indirectly, our own. Ecological data are indispensable for protecting nature: without measurements, we could not assess degradation, climate change, or the loss of biodiversity. But that data becomes more meaningful when it does not lead us to forget that we are speaking of a living web on which we depend.

Here, a dialogue with India's wisdom traditions can be enlightening, provided it is not distorted. The Upanishads and "Advaita Vedānta" do not anticipate molecular biology, do not explain how the brain functions, and do not offer alternative formulas to quantum physics. It would be a mistake to seek in them a hidden science ready to confirm our contemporary beliefs.

Their contribution, if anything, lies on another level. They challenge the image of an isolated self, standing before the world like an owner facing his possessions. They invite us to ask whether the separation we experience every day—I here, the rest out there—is truly the final word on our condition.

Schrödinger was deeply captivated by this question. His conviction that consciousness is not a multiplicity of completely separate entities belongs to his philosophical and personal worldview, not to a conclusion imposed by the equations of quantum mechanics.

To know without possessing means maintaining this sense of restraint even in human relationships. We can know a great deal about someone—their history, their psychological profile, their vulnerabilities—and yet not have truly encountered them. Sometimes we use explanations to shield ourselves from that encounter. Saying that a person is "merely" anxious, narcissistic, depressed, traumatized, brilliant, or fragile may contain a grain of truth. But it can also become a label that shuts the door. A human being never fully coincides with the description we— —make of them. This is not because they are a magical enigma beyond all understanding, but because they live in time, change, interpret themselves, surprise others, and sometimes even surprise themselves.

Think of a night spent keeping vigil beside a sick person. Medicine can describe vital signs, the course of a fever, the effect of a drug, and the neurochemical processes underlying fear and fatigue. All of this is necessary; in some cases, it can mean the difference between life and death. But none of these descriptions, on its own, fully captures what is happening in that room: the breath heard in the darkness, the wait for morning, the hand clasped in another's, the anguish and the care.

There are not two realities—one scientific and one human—that have nothing to say to one another. There are different levels of the same experience. The risk is not in explaining too much; it is in believing that a single type of explanation is authorized to speak for all the others. Knowledge as participation is therefore not an invitation to renounce reason. It is an invitation to make it more welcoming. To use maps without mistaking them for the landscape. To seek causes without forgetting meanings. To study life without losing the sense of our belonging to life.

Perhaps this is one of the most delicate legacies we can inherit from Schrödinger. Not the comforting idea that physics has already solved the mystery of consciousness, nor the opposite temptation to abandon rigor in favor of taking refuge in evocative words. Rather, the willingness to acknowledge that understanding does not mean grasping everything.

The value of silence.

There is a silence that is not the same as the absence of words. It is not the awkward silence of someone who doesn't know what to say, nor the silence imposed by authority on those who cannot speak. Rather, it is the silence that begins when we have pushed words to their limit and realize that, beyond that limit, there remains something that cannot be dismissed simply because we do not yet know how to articulate it.

Modern science was born, in part, out of a rebellion against sterile silences: those of superstition, dogma, and answers handed down by tradition. It has made matter speak through experiments, measurements, and public comparisons; it has learned to ask nature not what it wished to hear confirmed, but what the phenomena allowed it to say. And yet, even the most rigorous science knows another form of silence: that of suspending judgment, of waiting in the face of an unexpected result, of the caution that prevents us from turning an effective formula into a complete explanation of reality.

This is not a marginal approach. It is a prerequisite for serious research. When, on September 14, 2015, the two large LIGO interferometers in the United States—one in Hanford, Washington, and the other in Livingston, Louisiana—detected a tiny signal almost simultaneously, a discovery was not announced immediately. It should be noted that Italy is also home to the Virgo interferometer, located in Cascina near Pisa, but at that time it was undergoing an upgrade to *Advanced Virgo* and was not collecting data. Therefore, it was unable to record that event . The signal, later designated GW150914, was consistent with a gravitational wave produced by the merger of two black holes that had occurred approximately 1.3 billion years earlier. But between the data collection and the announcement, there were months of verification: cross-checks, independent analyses, ruling out possible instrumental noise, and comparing theoretical models with observations. It was not until February 11, 2016, that the LIGO-Virgo collaboration made the discovery public.

That interval was not a void between two statements, but a form of active silence. The scientific community remained silent—in the best sense of the term—because it knew that an extraordinary piece of data does not justify a hasty conclusion. Prudence did not diminish the excitement of the discovery; it made it reliable. Before telling the world that they had detected a vibration in spacetime predicted by Einstein a century earlier, it was necessary to distinguish a real signal from noise. In this sense, scientific silence is not a renunciation of knowledge. It is its most difficult exercise: not proclaiming what one wishes to be true before having sufficient grounds for it.

Theories allow us to predict phenomena, build instruments, and understand profound relationships between events that are extremely distant from one another. But their effectiveness does not give us the right to confuse language with the totality of what that language seeks to describe. We can come to know the laws governing a system's behavior with increasing precision without thereby possessing, once and for all, the ultimate meaning of its existence.

This distinction is particularly important when quantum physics enters our imagination. To say that, prior to measurement, a quantum system is described by a superposition of possibilities does not mean asserting that everything is indefinite in the vague and poetic sense of the term. It means something much more precise, and at the same time more limited: the equations allow for extremely accurate probabilistic predictions regarding the possible outcomes of a measurement. The next step—the one that claims to transform this description into an immediate theory of consciousness, freedom, or the soul—is no longer dictated by physics. It is an interpretation.

Here, silence becomes an intellectual virtue. Not because the big questions are off-limits, but because they deserve to be asked without distorting them with overly hasty answers.

The experience of pain, the quality of the red we see, the appearance of beauty, our duty toward another person, the meaning of death: we can certainly speak about these topics. We can do so through philosophy, poetry, art, prayer, and human conversation. But it would be naive to expect that a physical formula, however precise, could exhaust their lived meaning. Physics does not fail because it does not provide us with a complete answer regarding the value of an act of love or the meaning of suffering. That is not the task it has set for itself. We are the ones who fail when we ask it to be, all at once, cosmology, psychology, ethics, and consolation.

The wisdom traditions of Asia have developed, in various forms, a discipline of silence that can help us understand this point without turning it into a vague celebration of the ineffable. In those traditions, silence is not always the last word: it is often the first condition for listening more closely to the question.

In *the "Cūḷa-Māluṅkya Sutta,*" a text from the Pali Buddhist Canon also known as *"The Short Discourse to Māluṅkyaputta,*" a monk asks the Buddha to clarify a series of questions: Is the world eternal or not? Is it finite or infinite? Are the life force and the body the same thing or are they different? What happens to a fully awakened being after death? If he does not receive answers, he says, he will abandon the monastic life.

The Buddha does not resolve these questions with an alternative doctrine. Instead, he tells the parable of the man struck by a poisoned arrow. If this man were to refuse to have the arrow removed until he knew who had shot it, what caste the archer belonged to, what kind of wood had been used for the bow, and what kind of feathers for the arrow, he would die before obtaining his answers. The point is not that questions about the cosmos are illegitimate. It is that some questions, under certain circumstances, can become an escape from the most urgent issue: how to deal with the suffering that is already here, in real life.

What is called, in this case, "noble silence" is not indifference in the face of mystery. It is a form of discernment. Not everything that can be asked leads to greater clarity; not everything that is conceivable can be useful at the very moment we think it. The lesson does not ask us to stop questioning the world, but to recognize when the mind uses questions to delay confronting what truly concerns it.

Taoist thought also offers a valuable image. In Chapter 26 of the *Zhuangzi*, a Chinese text attributed to the tradition of Zhuang Zhou, we read: *"Words exist because of meaning; once the meaning is grasped, the words can be forgotten. Where will I find a man who has forgotten words, so that I may speak with him?"*

This image does not invite us to despise language. Without words, we would have no science, no historical memory, no rational discourse, and no transmission of knowledge. But words are like a net or a fish trap: necessary tools, not the reality they capture. The danger begins when we become attached to the net and forget the sea. It is a danger that concerns the mystic as much as the scientist, the philosopher as much as the reader of a popular science book . Those who abandon themselves to an indeterminate spiritual language may mistake emotion for knowledge. Those who possess a sophisticated technical language may mistake the ability to calculate for a complete understanding of existence. In both cases, what is lost is contact with reality: not reality reduced to a reassuring idea, but that which continues to resist our categories.

Thomas Stearns Eliot, in the first chorus of *"The* Rock" (1934), distilled this loss into two questions that have become famous: "Where is the wisdom we have lost in knowledge? Where is the knowledge we have lost in information?"

Today, these words resonate with perhaps even greater force than when they were first written. We possess a quantity of data that no other civilization has ever possessed. We can track the motion of the planets in real time, sequence genomes, simulate the evolution of complex systems, and observe galaxies billions of light-years away. Yet the accumulation of information does not automatically generate discernment, depth, or attention. An answer obtained in a matter of seconds does not necessarily correspond to a question that has been truly understood.

Silence, then, can be understood as a pause between knowledge and its assimilation. It is the moment when we ask not only, "What can I do with this knowledge?" but also, "What does it ask of me to change in the way I see things?" It does not take us away from reason; it takes reason away from haste. It does not reveal a truth hidden behind words; rather, it prevents us from treating words as if they were already the truth.

For Schrödinger, who spent his entire life seeking a unified vision of life and who looked with interest at the great insights of ancient India, this attitude would have held particular significance. It did not mean confusing physics with Vedānta, nor using the East as a spiritual ornament for scientific enigmas. It meant, more soberly, acknowledging that knowledge can become deeper when it loses the desire for absolute domination. The scientific mind, at its best, is not the mind that claims to have encapsulated the world in a formula. It is the mind that knows how to measure, distinguish, and correct itself; and which, when faced with what it cannot yet express, does not invent a certainty to fill the void.

Perhaps this is the most difficult lesson of silence: not to stifle the question, but to make it truer.

IX. Beyond the Paradox.

"The task is not so much to see what no one else has yet seen, but to think what no one has yet thought about what everyone sees."

Erwin Schrödinger, *Nature and the Greeks and Science and Humanism*, 1954.

"The world's greatest problems are the result of the difference between the way nature works and the way people think."

Gregory Bateson, *Mind and Nature: A Necessary Unity*, 1979.

"The best physical model for a cat is another cat, or, preferably, the cat itself."

Arturo Rosenblueth and Norbert Wiener, "The Role of Models in Science," Philosophy of Science, 1945.

Having reached the end of this journey, we can recognize that Schrödinger's cat is merely the best-known example of a much broader unease. Its true legacy lies neither in a textbook paradox nor in a formula capable of uniting quantum mechanics and spirituality. It lies in an uncomfortable question: Can we truly understand reality by separating matter from life, life from the mind, and the mind from the world it knows with such certainty?

Physics, biology, and the study of consciousness each have their own methods, languages, and questions. It would be irresponsible to abolish these differences in the name of a vague idea of unity. Yet, just as the sciences become more precise, the boundaries we have inherited appear less straightforward than they once seemed. An organism is not just any machine; a mind is not an external object like any other; the observer cannot be completely erased from the description of a world that he himself observes. "Beyond the paradox" does not, therefore, mean beyond reason. It means beyond the illusion that understanding always amounts to division. It means attempting to hold together the rigor of science and the depth of human questions, without turning either into a substitute religion.

Is a science possible that does not separate matter, life, and consciousness? Schrödinger's cat has become a proverbial creature. We encounter it in school textbooks, cartoons, movies, and conversations where quantum physics is invoked as a synonym for mystery. It is both alive and dead, they say; it is locked in a box; it waits for someone to look. But the paradox proposed by Schrödinger was not intended to make physics more enigmatic. On the contrary: he wanted to show how difficult it was to apply the rules of the atomic world to the objects of our everyday experience without caution.

That provocation continues to speak to us because it forces us to break a mental habit. We readily imagine the world as a collection of clearly separate things: on one hand, inert matter; on the other, living organisms; on one hand, the brain; on the other, consciousness; on one hand, the observing subject; on the other, a reality that would be fully defined even without him. These are useful distinctions, often necessary. Without them, we would have neither laboratories nor medicines nor reliable conceptual tools. But they can become cages when we forget that they are distinctions of thought before they are definitive divisions in reality.

Schrödinger did not offer a conclusive theory capable of healing all these rifts. Thanks to him, we do not possess a formula that demonstrates the unity of matter, life, and consciousness. To say otherwise would be to attribute to him what he was never able to prove. His importance lies elsewhere: in having refused to regard those rifts as obvious, definitive, or beyond question.

In 1944, he published **What Is Life?** The book was brief, written without the technical apparatus of a treatise, and yet it had an extraordinary impact. In it, Schrödinger posed a problem that was simple only in appearance: how can an organism maintain an ordered form over time, grow, repair itself, and transmit hereditary instructions, even though it is made of the same matter that, left to itself, tends to disperse its own differences?

Schrödinger's answer was not the final word on biology; in some respects, it was corrected or superseded by subsequent developments. His intuition regarding an "aperiodic crystal" as a possible material basis for heredity, however, evocatively anticipated the idea of a molecular structure capable of containing information without being a regular repetition like an ordinary crystal. Nine years later, in 1953, James Watson and Francis Crick would propose the double-helix model of DNA at Cambridge, drawing in part on the decisive X-ray diffraction data obtained by Rosalind Franklin and Raymond Gosling at King's College London.

Watson recalled on several occasions the impression Schrödinger's book had made on him. It would, however, be irresponsible to construct a linear narrative: **What Is Life?** did not "discover" DNA, nor did it provide the mechanism of molecular genetics. It did something different—and perhaps no less important—: it showed many young researchers that life could be investigated using physical tools without thereby being reduced to a mechanism devoid of history, form, and biological meaning.

Here it is worth pausing to consider an essential distinction. To say that an organism obeys the laws of physics is indisputable. Nothing living violates chemistry, thermodynamics, or the known laws of matter. But to say that life is merely physical—as if a complete description of particles rendered words such as development, adaptation, memory, vulnerability, nutrition, and behavior superfluous—is a further leap. It is not an inevitable experimental conclusion: it is a choice of language and perspective.

A violin is made of wood, varnish, glue, strings, and vibrating air. Knowing everything about the chemical properties of wood and acoustic frequencies does not make it unnecessary to speak of music, style, or interpretation. Not because music is a phantom that magically adds itself to matter, but because the organized whole produces a level of experience and meaning that does not coincide with the list of its materials. The same is true, with greater complexity, for an organism. Cells, molecules, and electrical signals are indispensable for describing it; yet they do not constitute the complete description of what happens when a seed becomes a tree, when an animal seeks its environment, or when a human being recalls the face of a loved one.

Barbara McClintock, an American geneticist and 1983 Nobel Prize laureate in Medicine, was well acquainted with the rigor of experimental work. For decades she observed corn, studying the form and behavior of chromosomes with almost obstinate patience. It was she who demonstrated the existence of transposable genetic elements—segments of DNA capable of changing position within the genome. At first, in the 1940s and 1950s, her ideas were met with skepticism; it was not until much later that their importance was fully recognized.

Yet McClintock did not describe her relationship with the organism as a mere collection of data. She told Evelyn Fox Keller: *"I feel that a person who works with living beings and seeks to understand them must develop an intimate, concrete, and practical knowledge of the living being under study.*" She did not ask scientists to abandon the scientific method in favor of intuition. She asked for something more demanding: an attentiveness capable of not immediately imposing on life the pattern we wish to find there.

This is a delicate point. Science does not gain depth by sentimentally humanizing cells or attributing intentions to a molecule. But it can become impoverished when it views a living being as if it were merely a passive object, devoid of its own coherence. An organism is not a highly complex rock. It exchanges matter and energy with its environment, maintains its own organization for a certain time, responds to disturbances, develops, in many cases reproduces, and changes over the course of evolution. Even in the most elementary forms of life, we encounter a dynamic that is not merely the assembly of parts.

In the 1970s, Chilean biologists Humberto Maturana and Francisco Varela proposed the term autopoiesis, from the Greek: *"self-production."* With this word, they referred to a fundamental characteristic of living beings: they are not merely products of external forces, but continuously maintain and renew their own organization through a network of processes. An organism, of course, depends on its environment; it breathes, feeds, receives stimuli, and suffers injuries and constraints. But it cannot be reduced to the sum of what the environment does to it. It responds, selects, transforms, and preserves a certain identity through change.

This idea does not solve the mystery of life, and it should not be used as a magic word. It does, however, help us see why the boundary between matter and life is not merely a matter of quantity. It is not enough to bring together enough atoms to automatically produce an organism. A form of organization is required that is capable of sustaining itself over time, of transforming matter into a living process.

The philosopher and anthropologist Gregory Bateson had sensed just how dangerous our mental divisions could become when they do not respect the fabric of real relationships. He wrote: *"The world's greatest problems are the result of the difference between the way nature works and the way people think."* Bateson was no enemy of scientific analysis. He had worked in anthropology, psychiatry, biology, and communication theory. His target was the tendency to isolate what, in real life, exists only in relation to other things.

A forest is not merely the sum of its trees. It also includes the soil, the rainfall patterns, the fungi that live in association with the roots, the insects, the bacteria, the animals, the climate, the slow transformations, and the sudden events. Separating these elements to study them may be necessary; but forgetting that they influence one another can lead to serious errors. The same applies to the human body. We can study a neuron, a gene, a hormone, or a region of the brain. But none of these elements, isolated from their context, is equivalent to a living person.

This is where the problem of consciousness resurfaces, without claiming a definitive solution. We know much more about the nervous system than we did a century ago: we can observe brain activity and identify circuits involved in perception, language, memory, and pain. And yet the simplest and most difficult question remains open: why are certain physical processes accompanied by lived experience? Why are there not only electrical signals and chemical reactions, but also the sensation of pain, the taste of coffee, the color of a sunset, the weight of a memory?

Here, science has the final word: there are strong empirical reasons to believe that consciousness, as we know it, depends closely on the activity of the brain and the body. Injuries, anesthesia, medications, sleep, and neurological diseases profoundly alter experience. To ignore these facts in the name of a disembodied spirituality would be an escape from reality.

Here, however, philosophical interpretation comes into play: the interdependence between experience and the brain does not automatically imply that experience is an illusion or a negligible detail. Explaining the conditions that make an experience possible is not yet the same as saying what it means to be within it. The map of the brain areas active when we listen to Bach is not the act of listening to Bach itself. It is valuable knowledge of what happens in the brain during that listening; it does not replace the music, nor the person who is moved by it.

We do not possess a body the way one possesses an object. We are corporeal. We see through eyes situated in a head , we touch with hands that can themselves be touched, we inhabit a space that is not an abstract diagram but a field of possibilities, obstacles, distances, and encounters. Consciousness does not seem to arise in an immaterial void; it manifests itself in an embodied life, exposed to time and the world.

Schrödinger, with his idea of a consciousness that is not absolutely multiplied, went further than most scientists of his time. Drawing also on the Upanishads and "Advaita Vedānta," he found there the image of a profound unity behind the plurality of individual perspectives.

What do we mean when we say "I"? An entity completely closed in on itself, separated from the body, from history, from others, and from nature? Or a point of view that is certainly unique, yet formed from the very beginning within a network of relationships that precedes and sustains it?

The Indian tradition offers its own language for this question, one that cannot be equated with biology or physics. When certain Upanishads speak of the identity between Ātman and Brahman, they are neither anticipating field theory nor describing the brain in ancient terms. They are questioning the experience of separation: they ask whether the "I" that feels isolated truly exhausts what we are. Their proposal may be accepted, debated, or rejected; but it must not be trivialized.

The unity we can speak of seriously today is, first and foremost, a unity of relationship. We are made of the same matter as the stars, but this phrase, though poetically evocative, is not enough to explain the life of the mind. We are organisms that have emerged from a long evolutionary history, but even this does not fully capture what happens when we ask a question about our place in the universe. We are bodies, biological systems, social beings, creatures capable of f language, and of self-awareness. None of these levels negates the others.

Norbert Wiener, an American mathematician and one of the founders of cybernetics, wrote with the Mexican physiologist Arturo Rosenblueth that *"the best material model for a cat is another cat, or, preferably, the cat itself."* The quip, published in 1945, is more serious than it seems. It reminds us that, to fully reproduce the complexity of a living being, we may need the living being itself. Models are indispensable: they allow us to simplify, predict, and verify. But the model is never the animal that sleeps in the sun, avoids danger, seeks food, feels pain, follows a scent, ages, and dies.

This gap between the model and the living being does not represent a failure of knowledge. It represents its fruitful limit. Science advances precisely because its models are not confused with the world: they are put to the test, refined, and replaced when they no longer suffice. The problem arises when a description that is effective in one field claims to become the only legitimate language for everything that exists.

A science capable of not separating matter, life, and consciousness should therefore not be a "super-science" that absorbs every discipline and claims to have the final word. It should be, more modestly and more courageously, a science aware of its own limits and open to dialogue. Physics can offer biology essential tools; biology can show philosophy that the mind is inseparable from the body's vulnerability; philosophy can help the sciences clarify the words they use—reality, explanation, information, observation, subject—before they become slogans.

This dialogue does not necessarily produce peace. Sometimes it produces conflict, and conflict can be useful if it is honest. Schrödinger remains relevant precisely because he cannot be easily categorized. He was one of the fathers of wave mechanics but was also a restless thinker, unwilling to confine his questions within the reassuring limits of specialization. Some of his insights into life were fruitful; some of his reflections on the mind remain controversial; his affinity for Vedantic thought cannot be transformed into a scientific theory.

Yet Schrödinger's marginality is perhaps part of his legacy. There are thinkers who found a school of thought; others keep a wound open in thought. Schrödinger belongs more to the latter category. He forces us to ask whether our image of the world is broad enough to contain—without reducing them—the precision of physics, the complexity of life, and the immediate evidence of conscious experience. That is no small thing.

To inhabit unity without simplifying it means renouncing two opposing consolations. The first is hasty reductionism: the idea that once the components have been explained, there is nothing left to understand. The second is facile mysticism: the idea that simply uttering words like "energy," "quantum," or "universal consciousness" is enough to resolve real difficulties. Between these two shortcuts lies a more demanding path.

We can follow it without claiming to arrive at a definitive doctrine. Matter is not an enemy of the spirit; biology is not a degradation of philosophy; consciousness is not an insignificant accident in a universe devoid of inner life. But none of these statements, on its own, is a proof. They are invitations to continue the work of thought with greater precision and greater humility.

Perhaps Schrödinger's most enduring legacy is not the answer to a question, but the courage not to accept too quickly the divisions with which we try to make the world more manageable. The cat finally emerges from the box; but we are left facing a reality vaster than our categories, called not to possess it, but to observe it more carefully.

The Physicist Who Asked Too Many Questions.

There are thinkers who become famous for a formula, an image, or an invention. And there are thinkers whose fame ends up obscuring precisely what they had most important to say. Erwin Schrödinger belongs, in an almost paradoxical way, to both categories.

His name has entered the collective imagination thanks to an imaginary cat trapped in a box. It is a peculiar fate for a man who disliked oversimplifications and who had derived one of the fundamental equations of 20th-century physics. Schrödinger's equation, published in 1926, remains an essential tool for describing the behavior of quantum systems. Without it, we would not understand the structure of atoms, chemical bonds, or the functioning of semiconductors—a crucial part of the technologies that shape our daily lives.

Yet, if we were to ask a contemporary physicist, biologist, or philosopher what Schrödinger's true legacy is, the answers would quickly diverge. For some, he is one of the founders of quantum mechanics. For others, he is the author of **What Is Life?**, the short book that helped guide a generation of young scientists toward the mystery of biological heredity. For still others, he is an unconventional thinker, difficult to categorize—too close to philosophy to be accepted without reservation by specialized science, and too rigorous in his scientific training to be easily assimilated into any philosophical school of thought.

This difficulty in categorizing him is not a mere biographical detail. It is a sign of a body of thought that remains relevant precisely because it refuses to be tamed.

Schrödinger remains marginal, not in the sense that he has been forgotten—it would be absurd to say so. His name is everywhere. He remains on the margins in a more subtle sense: his broader questions are often treated as a private aside, almost an eccentricity to be carefully distinguished from his scientific work. The physicist is accepted; the thinker who dares to question the relationship between matter, life, and consciousness is viewed with a certain caution. It is as if one were to say: let's take the equation; let's leave the rest aside.

But for Schrödinger, the rest was no mere pastime. In 1943, he was in Dublin, where he had accepted an invitation from Éamon de Valera, then head of the Irish government and founder of the *"Dublin Institute for Advanced Studies."* Europe was at war. Many laboratories were directly or indirectly involved in the war effort; physics was demonstrating its destructive power with terrifying clarity. In that situation, Schrödinger chose to devote a series of public lectures not to weapons, not to elementary particles, but to a seemingly simple question: what distinguishes a living organism from inanimate matter?

He was not searching for a soul hidden within the cell. He did not propose to exempt biology from the laws of physics. On the contrary, he wanted to understand how life could adhere to those laws and, at the same time, produce order, continuity, form, and hereditary memory. The language of **What Is Life?** retains some inevitable limitations: molecular genetics had not yet emerged, the double-helix structure of DNA would not be proposed by James Watson and Francis Crick until 1953, and many of

Schrödinger's statements on entropy were later refined. But the question he posed remained fruitful.

Schrödinger had not discovered DNA, nor had he predicted the future of genetics in detail. He had done something different—and perhaps no less remarkable: he had identified a point where distant disciplines were bound to converge.

His merit did not lie in already having the answer. It lay in having seen that the question could no longer be confined to a single room.

Today we are accustomed to speaking of dialogue between physics, biology, neuroscience, computer science, and philosophy. But this habit risks misleading us. The disciplines do engage in dialogue, certainly, but they often maintain very rigid boundaries. Each protects its own vocabulary, its own methods, its own evidence. It is right that they do so: without rigor, dialogue quickly turns into confusion. And yet there is another, less obvious danger: that of dividing reality into territories so separate that we can no longer understand how they fit together.

Matter is entrusted to physics. Life to biology. Conscious experience to neuroscience or psychology. Meaning to philosophy, art, and religion. Then, once the responsibilities have been assigned, we act as if the problem were solved. Schrödinger suspected that it was not solved at all.

The uncomfortable aspect of his thinking does not lie in proposing a ready-made solution. It lies in rejecting the idea that we can truly understand anything by completely separating the world we study from the experience of those who study it. Science, of course, must distinguish: it measures, compares, isolates variables, and constructs models. That is its strength. But this methodological distinction can, almost without our noticing it, transform into a broader conviction: the idea that what is separable in the laboratory is also separate in reality.

Here we must proceed with caution. To say that matter, life, and consciousness should not be conceived as c r entirely separate worlds does not mean asserting that a stone thinks, nor that quantum physics proves ancient spiritual doctrines. These shortcuts have the advantage of being evocative and the drawback of explaining almost nothing. Schrödinger himself, though drawn to the Indian tradition and its idea of a fundamental unity of reality, possessed no physical proof of that vision. When he stepped outside the realm of equations, he entered that of philosophical reflection. To confuse the two planes would be to betray the very caution that his work demands.

But recognizing this distinction does not require us to ignore the questions he raised. One of them concerns consciousness. For much of modern science, consciousness has been treated as a secondary phenomenon: important for human life, certainly, but not fundamental to the description of the world. Equations work without needing to speak of joy, pain, color, fear, memory, or presence. And this is perfectly understandable: a formula does not have to describe how it feels to look at the sea or to lose a loved one.

Yet, everything we call knowledge takes place within the context of an experience. No theory exists on its own, outside of any mind, on an empty table in the universe. There are tools, data, calculations, research communities, and shared languages; but, at some point, there is always someone who observes, interprets, corrects, understands, or fails to understand.

Francisco Varela, who was trained many decades after Schrödinger and was as attentive to the cognitive sciences as he was to Buddhist contemplative practices, coined a phrase that could well capture this unease: *"The mind is not in the head*." He did not mean to deny the brain. He wanted to remind us that the mind cannot be understood as an isolated object, enclosed within the skull, because it exists within the relationship between a body, an environment, other bodies, a history, and a language.

This insight differs greatly from traditional metaphysics, yet it retains something of Schrödinger's own unease: the human being is not an external observer who, from a privileged vantage point, gazes upon a universe that is already complete and finished. The human being is a part of the world attempting to understand the world. And this situation is not an accident to be eliminated; it is the starting point from which all knowledge begins.

Perhaps this is also why Schrödinger continues to be received with caution. Contemporary scientific culture, especially in its most effective and specialized forms, tends to reward well-defined questions. Which protein regulates a certain process? Which neural circuit is involved in a choice? Which mathematical model describes a given phenomenon? These are indispensable questions. They produce verifiable results, therapies, tools, and predictions.

But questions like "What is life?" or "How can matter give rise to experience?" are of a different nature. Not because they are vague, but because they are too vast to belong permanently to a single field. They force us to slow down. They call into question the language we use to categorize problems. They ask us to acknowledge that, even after collecting a great deal of data, we may still not have grasped the meaning of the whole.

The French writer and physician Georges Canguilhem, a leading scholar of the history of the life sciences, observed that life is not merely an object to be measured: it is also the capacity to establish values, to distinguish what fosters existence from what threatens it. An organism is not a machine that blindly executes a program; it is an exposed, vulnerable being, capable of adapting, failing, healing, and inventing new forms of balance.

This perspective does not replace molecular biology, but it prevents us from believing that living beings are fully accounted for by their chemical description. Similarly, talking about neurons and synapses does not render superfluous the question of what it means to feel, to choose, to suffer, or to be present to oneself. Explanations of mechanisms are necessary; however, they are not always sufficient to tell us what we are looking at.

Schrödinger therefore remains relevant today because our era has made his questions even more urgent. We know how to manipulate the genome, build artificial intelligence systems capable of producing language, and observe brain activity with a precision that was unthinkable in 1943. And yet we have not untangled the fundamental knot: how can we hold together what we know about matter, what we recognize in life, and what we experience in consciousness without reducing one level to the other?

The temptation to reduce things is strong, because it gives the impression of mastery. If life is merely chemistry, chemistry merely physics, and consciousness merely brain activity, it seems as though the world is finally organized into a simple hierarchy. But that "merely" is often where something essential is lost. Not because chemistry is false, or physics insufficient, or neuroscience useless—on the contrary, because each of these fields of knowledge is too valuable to be reduced to a totalizing philosophy.

Schrödinger's unsettling idea may lie in this: reality does not ask us to choose between scientific precision and philosophical depth. It asks us not to use one to silence the other.

Toward the end of his life, having returned to Vienna in 1956 after years spent in Ireland, Schrödinger remained a celebrated yet reclusive figure. He had received the Nobel

Prize in 1933, shared with Paul Dirac; he had helped transform physics; he had lived through some of the century's greatest intellectual and political upheavals. Yet he never ceased to look beyond his own discipline. Not because he despised science, but because he took it too seriously to imagine it as a catalog of definitive answers.

Perhaps his true legacy is not a doctrine to be accepted. It is a restlessness that must not be dismissed. The idea that matter is not a simple concept, that life is not a negligible accident, and that consciousness cannot be treated merely as a side effect of impersonal processes.

We do not yet know what language will be able to hold these aspects together without forcing them. But we can acknowledge that the question remains open. And that, precisely for this reason, it continues to be alive.

The cell, the atom, and the question that unites them.

Having acknowledged how Schrödinger's thought has, in part, remained on the margins of mainstream scientific culture, we can ask ourselves what it truly asks of us. Not to abolish the boundaries between disciplines, not to transform physics into a form of spirituality, nor to seek confirmation of great philosophical insights in biology. It asks for something more sober and, perhaps, more difficult: to build bridges without pretending that the shores are the same thing.

Physics, biology, and philosophy speak different languages because they observe different aspects of reality. Physics seeks regularities, structures, and measurable relationships. Biology studies living beings in terms of their organization, their evolutionary history, and their capacity to transform and sustain themselves. Philosophy questions the meanings of the words we use, the limits of explanations, and the place of human experience within the framework we construct.

The problem does not arise from this difference. It arises when each language claims to be sufficient in and of itself. An organism, for example, is certainly composed of atoms. And atoms obey the laws of physics. But saying this does not yet amount to understanding what an organism is. A living cell is not simply a small container of molecules. It is a system that exchanges matter and energy with its environment, repairs certain damage, maintains a certain continuity, reacts, changes, and sometimes reproduces. A forest is not merely the sum of its trees; a body is not merely the sum of its organs; a person is not merely the sum of the processes occurring in their neurons.

This is not about denying the reality of matter. It is about not confusing knowledge of the components with an understanding of the whole.

The Scottish biologist and mathematician D'Arcy Wentworth Thompson, in his seminal book **On Growth and* Form*, published in 1917, formulated an image that still retains a surprising power today: *"The form of an object is a 'diagram of forces.'"* Thompson observed shells, bones, horns, the structures of jellyfish, and the geometries of plants with the patience of a naturalist. He saw in living form not a random ornament, but the visible result of tensions, constraints, movements, and interactions with the environment.

His insight was not a poetic vision of nature. It was an invitation to look more closely. The shape of a bone, for example, tells the story of the weight it has borne; the spiral of a shell records a growth rhythm; the arrangement of leaves around a stem responds to issues of light and space. Life does not escape the laws of physics: it traverses them, utilizes them, and organizes them into configurations that we cannot adequately describe by resorting to a single scale of observation.

Here, the dialogue between physics and biology becomes essential. Physics can show which conditions make a form possible; biology asks why that form has been preserved, modified, or abandoned in the course of evolution. The former studies the possibilities of matter; the latter recounts the adventures of matter when it becomes alive.

Schrödinger had clearly sensed this need. He did not imagine that he could solve the problem of life with some concept borrowed from physical ph . Rather, he sought to understand how matter could maintain, within organisms, such a delicate yet resilient form of order.

This was a question posed before the discovery of the structure of DNA. But even after that discovery, Schrödinger's question was not truly resolved. Knowing where and how a portion of hereditary information is stored still does not explain what turns a set of chemical instructions into a living being.

The geneticist and Nobel laureate François Jacob put it in a brief, almost disarming sentence: *"The dream of every cell is to become two cells.*" The phrase appears in **The Logic of* Life*, published in French in 1970. It is a memorable remark not because it naively humanizes the cell, but because it reveals an essential property of life: the living tends to continue, to duplicate a form, to pass on a history.

Of course, a cell does not dream in the sense that we do. It has no nighttime images, autobiographical memories, fears, or plans. But Jacob's formula helps us distinguish between two levels. On the one hand, there are molecular reactions, which can be described with great precision. On the other hand, there is the dynamic unity that these reactions make possible: a cell that lives, feeds, defends itself, divides, or dies. Biology does not abandon chemistry; it asks chemistry to participate in a broader narrative.

This is where philosophy can play a valuable role. It does not add experimental data to what is gathered in laboratories. It does not replace microscopes, genetic sequencers, or equations. But it asks what we mean when we use words like "life," "organism," "information," "mind," "cause," and "order." Words that seem familiar, but become less obvious as soon as we try to define them.

Take the concept of information. In genetics, talking about information is both useful and inevitable. DNA is transcribed, translated, and copied; it contains sequences that contribute to the construction of proteins. But the language of information can mislead us if we imagine the gene as a sovereign command and the organism as an obedient machine. Genes always act within a network of relationships: with other genetic sequences, with cells, with the body's internal environment, with nutrition, and with the history of development.

The American biologist Richard Lewontin, in his 2000 book **The Triple* Helix*, put it clearly: *"Organisms do not find the world in which they develop. They construct it."* This does not mean that an organism can invent reality as it pleases. A beaver does not create a river out of thin air, and a human being does not determine the laws of gravity. But the beaver builds dams that alter the course of the water, creating a new environment for itself and for other species. Living beings are not merely shaped by the world; they help shape the world in which they live.

This observation has implications that extend beyond biology. We, too, inhabit environments that are not only natural but also linguistic, social, technological, and symbolic. Cities, schools, hospitals, digital networks, eating habits, the images we use to represent the body and death—all of this shapes the way we live, perceive, and even fall ill. Matter does not disappear, but enters into a web of meanings.

At this point, it is essential to maintain a crucial distinction. Quantum physics does not prove that consciousness creates the world. It does not prove that every living being is part of a cosmic mind. It does not prove the Upanishads, nor does it automatically confirm "Advaita Vedānta." Yet it would be just as hasty to conclude that science has nothing to say to philosophy and that philosophy has nothing to ask of science. Great scientific revolutions have always transformed the image that human beings have of themselves.

When Copernicus published *_De revolutionibus orbium_ coelestium* in Nuremberg in 1543, he did not merely propose a new astronomical model. Over time, that model helped shift humanity away from the immovable center of the universe. When Charles Darwin published *_On the Origin of_ Species* in 1859, he did not merely provide a theory of evolution: he made it more difficult to conceive of life as a fixed hierarchy of beings ordered once and for all. And when quantum physics challenged the image of a world composed of objects that always possess definite properties, independent of any experimental context, it did not merely change a page in physics textbooks. It challenged our need to imagine reality as a simple scene, already fully laid out before us. Alfred North Whitehead, an English mathematician and philosopher who had collaborated with Bertrand Russell before moving to the United States, wrote in 1925, in *_Science and the Modern World_*: _"Nature is a structure of evolving processes. Reality is the process."_ This is a statement that must be handled with care. It does not replace contemporary physics, nor does it provide us with a definitive theory of the cosmos. But it contains an important warning: perhaps we are accustomed to thinking of the world as a collection of things, whereas many of its deepest realities are best understood as relationships, transformations, and transitions.

A river is not a static thing, and yet it is a river. A body is constantly renewing its cells; it breathes, excretes, assimilates, and ages; and yet, we recognize a continuity. A person goes through seasons, losses, encounters, and changes in character; and yet they are not simply replaced by another person every morning. Life seems to teach us that identity does not always coincide with immobility.

Lynn Margulis, an American biologist best known for her work on endosymbiosis (the now widely accepted idea that certain fundamental parts of complex cells derive from ancient forms of cooperation between different organisms), liked to say: *"Life did not conquer the globe through combat, but through networks.*" This quote, from **Symbiotic* Planet*, does not deny evolutionary competition. Margulis was no nature romantic: she knew full well that life involves predation, conflict, and extinction. But she wanted to correct an overly one-sided view. Evolution is not just a struggle of all against all; it is also about association, mutual dependence, and the invention of new living units arising from unlikely encounters.

We ourselves are an example of this. Mitochondria—the cellular structures involved in energy production within our cells—retain traces of an ancient bacterial origin. In the distant past, an organism incorporated a bacterium without destroying it; from that symbiosis, which lasted millions of years and was transformed by evolution, a new form of cellular life emerged. Our body therefore carries, in its most basic functioning, the memory of an ancient collaboration.

This is not a symbol: it is biology. But it can also become an opportunity for reflection. It reminds us that the individual is not always a closed fortress. Sometimes it is a stabilized community, a relationship that has become a body. No single perspective encompasses the totality of reality. Every living being encounters the world according to the possibilities and limitations of its own body. Even human beings. And here, once again, philosophy and science can meet without becoming confused. Science helps us understand the mechanisms of perception; philosophy prevents us from forgetting that knowing never means passively receiving a world that is already ready-made. It means entering into a relationship with it—through a body, a language, a history, a community of other observers. Perhaps the most important bridge is not the one that attempts to reduce everything to a single language. It is the one that makes a rigorous conversation between different languages possible. Physics teaches us that the world is stranger and less intuitive than our everyday experience suggests. Biology shows us that life is not a mere mass of matter, but an organized, vulnerable, creative form capable of transforming its environment. Philosophy reminds us that no description, however powerful, eliminates the question of the meaning of what we describe.

Schrödinger did not leave us a complete map for navigating these territories. Rather, he left us an intellectual gesture: the refusal to dismiss as irrelevant the questions that arise at the boundaries of disciplines. It is a gesture that requires caution, because bridges can be fragile and analogies misleading. But it also requires courage, because knowledge that does not communicate with other forms of knowledge risks becoming increasingly precise yet increasingly incapable of finding its bearings. We must not seek a formula capable of explaining atoms, cells, consciousness, and meaning all at once. Such a formula, if it were ever possible, would not suffice to encompass the full range of human experience. We can, however, learn not to treat these aspects as if they belonged to universes unrelated to one another. Perhaps this is the most sober and ambitious task: not to erase the differences, but to listen to what happens when matter becomes life, when life becomes experience, and when experience begins to question the mystery from which it springs.

Bibliography.

Aguirre Anthony	Zen e multiversi
Al-Khalili	La fisica del diavolo
Allen Steve	Inganni della logica
Ambesi Alberto	Il panteismo
Amir Dan Aczel	L'equazione di Dio
Amir Dan Aczel	Il mistero dell'alef
Amir Dan Aczel	Entanglement..
Anderl Sybilla	L' universo e io.
Andreoli Vittorino	L'origine della coscienza
Anthes Emily	Guida rapida per cervelloni: la mente
Arecchi Tito & Eva	I simboli e la realtà
Argentieri Niccolò	Ci sono elettroni nel mondo della vita?
Armstrong David	Che cos'è la metafisica
Asimov Isaac	L'universo invisibile
Asimov Isaac	In principio.
Asimov Isaac	Il libro della fisica
Baker Staphen	Il potere segreto dei matematici
BakerJoanne	Fisica quantistica. 50 grandi idee
Barbillini Amidei	Le domande di tutti
Barbour Julian	Il punto di Giano
Barbour Julian	La fine del tempo. Einaudi.
Barone Vincenzo	Albert Einstein. Il costruttore di universi
Barrow Tipler	Il principio antropico.
Barrow John David	Il mondo dentro il mondo
Barrow John David	Dall'Io al Cosmo
Barrow John David	L'universo come opera d'arte
Barrow John David	Il libro degli universi.
Barrow John David	L'infinito
Barrow John David	1+1 non fa (sempre) 2
Barrow John David	I numeri dell'universo
Barrow John David	Da zero a infinito ..
Barrow John David	Perché il mondo è matematico?
Barrow John David	Le Origini dell'universo
Barrow John David	Le immagini della scienza

Battiston Roberto	La prima alba del cosmo.
Beitman Bernard	I messaggi delle coincidenze.
Bellone Enrico	I corpi e le cose
Benini Arnaldo	Neurobiologia della volontà
Bergson Henri	Saggio sui dati immediati della coscienza
Berkeley George	Opere filosofiche
Bersani- Peres	Matematica proverbiale
Bersani R, Peres E.	Matematica. Corso di sopravvivenza
Berti Enrico	Storia della metafisica
Bocchi G, Ceruti M.	Origini di storie
Bodanis David	E=mc². Biografia dell'equazione che ha cambiato il mondo
Bodel-Tonelli	La nuova fisica delle particelle e i segreti dell'universo
Bogdanov Igor	Prima del Big Bang.
Bohm - Khrisnamurti	Dove il tempo finisce
Bohm - Khrisnamurti	I limiti del pensiero
Bōhm Conrad	Le chiavi del cosmo
Bohm David	Sul dialogo
Bojowald Martin	Prima del Big Bang.
Bonaventura, Colombo, Miluzio	L' universo su misura. Viaggio nelle incredibili coincidenze cosmiche che ci permettono di essere qui
Boncinelli Edoardo	Gli enigmi del tempo
Boncinelli, Ereditato	Tutto si trasforma. Breve storia dell'energia
Bui Daniele, LeoniSilvio	Mente, cervello e coscienza.
Burgin Luc	Errori della scienza
Byung-Chul Han	Le non cose.
Cambray Joseph	Sincronicità.
Campbell, Joseph	Le distese interiori del cosmo.
Campbell, Joseph	Dee. I misteri del divino femminile
Cantalupi - Santarcangelo	Psiche e realtà
Cantucci Pierluigi	La rivoluzione dell'intelligenza artificiale
Cappelletti Valentina	Dall'ordine alle cose. Saggio su Werner Heisenberg
Capra F. Luisi P.L.	Vita e natura. Una visione sistemica.
Capra Fritjof	La scienza della vita
Capra Fritjof	Verso una nuova saggezza
Capra Fritjof	Il Tao della fisica.
Capra Fritjof	Il punto di svolta
Carlson W. Bernard	Tesla. L'inventore dell'era elettrica

Casati Roberto	La scoperta dell'ombra
Castorina Paolo	La «particella di Dio» e l'origine della massa
Cattabiani Alfredo	Planetario
Cederquist John	Le coincidenze non esistono.
Centini Massimo	Il megalisismo. Luoghi sacri e di potere
Cercato Adriana	La scatola di luce
Cesati Cassin Marco	Non siamo qui per caso.
Chalmers David	Che cos'è la coscienza?
Chalmers David	Più realtà. I mondi virtuali e i problemi della filosofia
Chandrasekhar Sub.	Verità e bellezza.
Chiaberge Riccardo	La variabile Dio
Chinnici Giorgio	Guarda caso. I meccanismi segreti del mondo quantistico
Chopra Deepak	Le coincidenze. Sperling & Kupfer.
Ciuffardi, Perissi ok	La fine della coscienza? Dalla mente bicamerale alla intelligenza artificiale
Clark Michael	I paradossi dalla A alla Z
Close Frank	Teorie del tutto
Colpi Monica	Buchi neri evanescenti
Couliano Joan	I viaggi dell'anima
Crespi Marco	L'equazione della coscienza
Currivan Jude	L'ologramma cosmico
D'Agostini (Manuali)	Illuminismo
D'Ancona Vittorino	Mitologia greca
Dal Buono Lucio	Dall'atomo all'anima.
Damasio Antonio	Emozione e coscienza
Damasio Antonio	Sentire e conoscere
Damasio Antonio	L' errore di Cartesio.
Danielsson Ulf	Il mondo in sé. La coscienza e il tutto nella fisica
Dask Mike	Al di là dei confini
Davies Paul	La mente di Dio
Davies Paul	L'universo che fugge
Davies Paul	Il cosmo intelligente
Davies Paul	Dio e la nuova fisica
Davies Paul	Come costruire una macchina del tempo
Davies Paul	Siamo soli?
Davies Paul	Gli ultimi tre minuti
Davies Paul	Sull'orlo dell'infinito.
Davies Paul	I misteri del tempo.
Dawkins Richard	L'orologiaio cieco
De Agostini	Biologia

De Crescenzo Luciano	Ordine e disordine
De Crescnzo Luciano	Il tempo e la felicità
De Santillana Giorgio Von Dechend Hertha	Il mulino di Amleto.
Dehaene Stanislas	Cervello e coscienza
Delmastro Marco	Particelle familiari. Le avventure della fisica e del bosone di Higgs.
Dennett Caruso	A ognuno quello che si merita
Dennett David	Coscienza. Che cosa è
Di Francesco Michele, Marraffa, Tomasetta	Filosofia della mente.
Dizionario Mondadori	Dizionario della saggezza orientale
Donà Massimo	Magia e filosofia
Du Sautou Marcus	L'equazione da un milione di dollari
Du Sautou Marcus	Il disordine perfetto
Du Sautou Marcus	Le teorie impossibili.
Dumezil Geoges	La religione romana arcaica
Edelmen, Tononi	Un universo di coscienza.
Einstein Albert	Albert Einstein. Relatività e nuova fisica
Einstein Albert	Pensieri, idee, opinioni
Einstein Albert	Il mondo come io lo vedo
Einsten, Infeld	L'evoluzione della fisica
Ekelan Ivar	Il caos
Ereditato Antonio	Le particelle elementari
Fabbro Franco	I fondamenti biologici della filosofia.
Fabbro Franco Astrolabio Ubaldini	Neuroscienze e spiritualità.
Facchini Fiorenzo	L'avventura dell'uomo. Caso o progetto?
Facchini Fiorenzo	Origini dell'uomo e evoluzione culturale
Faggin Federico	Silicio
Faggin Federico	Irriducibile, La coscienza, la vita, i computer e la nostra natura
Falletta Nicholas	Il libro dei paradossi
Ferrini Federico	Le onde gravitazionali. Una nuova porta sul cosmo
Ferris Timothy	L' avventura dell'universo. Da Aristotele alla teoria dei quanti.
Ferrone Vincenzo	Una scienza per l'uomo
Feynman Richard	Sei pezzi facili

Feynman Richard	QED
Feynman Richard	Il senso delle cose
Fiscaletti Davide	Il quadro olografico. Le frontiere non locali della fisica moderna
Ford Kenneth	Il mondo dei quanti.
Forno Guido	Una memoria di ferro
Fosar-Bludorf	Ipercoscienza. L'intelligenza nasosta nel DNA
Foster Charles	Noi esseri umani. 40.000 anni di storia della coscienza
Franchi E.M.	La teoria dei tipi psicologici di Carl Jung
Fraser J. T.	Il tempo: una presenza sconosciuta
Frova Andrea	Luce. Una storia da Pitagora ad oggi
Frova Andrea	Il cosmo e il Buondio. Dialogo su astronomia, evoluzione e mito
Frova Andrea	Perché accade ciò che accade
Galfard Cristophe	Come capire E=mc2
Galilei Galileo	Dialogo dei massimi sistemi
Gallippi Angelo	Federico Faggin. Il padre del microprocessore
Gamow George	Gravità
Gamow George	Nuovo Mondo di mr. Tompkins
Gamow George	Trent'anni che sconvolsero la fisica
Gamow George	Le avventure di mr. Tompkins.
Gasperini Maurizio	L' universo prima del big bang. Cosmologia e teoria delle stringhe
Gazzaniga Michael	L coscienza è un istinto.
Giacomelli Giorgio	Dal quark al big bang
Giuliodori Lucio	Tra fisica e metafisica. Alcune implicazioni filosofiche della meccanica quantistica
Giuliodori Lucio	Sul concetto di sincronicità: Jung tra psicanalisi e quantismo
Gleick James	Caos. La nascita di una nuova scienza
Goff Philip	L'errore di Galileo. Fondamenti per un nuovo studio della coscienza
Goswami Amit	Evoluzione creativa.
Goswami Amit	Guida quantica all'illuminazione.
Graves Robert	I mito greci
GrazianoMichael	Ripensare la coscienza
Grecchi Vittorio	Il campo cervello mente
Greco Pietro	L' origine dell'universo
Greene Brian	Fino alla fine del tempo.
Greene Brian	La realtà nascosta
Greene Brian	L'Universo elegante
Greene Brian	La trama del cosmo. Spazio, realtà.

Gribaudo	Simply Meccanica quantistica
Gribbn John	Dizionario enciclopedico di fisica quantistica
Guido Paolo	Da Buddha a Einstein
Hack Margherita Ferreri Cossard	Il lungo racconto dell'origine.
Hall Nina	Caos. Una scienza per il mondo reale
Hand David	Il caso non esiste.
Harner Michael	La via dello sciamano
Harrison Edward	Le maschere dell'universo
Hawking - Penrose	La natura dello spazio e del tempo
Hawking Heckler	Il grande disegno.
Hawking -Mlodinov	Dal Big Bang ai buchi neri. Breve storia dell'universo. Mondadori
Hawking Stephan	Buchi neri e universi neonati
Hawking Stephen	La teoria del tutto. Origine e destino dell'universo. Mondadori.
Hawking Stephen	La grande storia del tempo. Mondadori
Hawking Stephen	L'universo in un guscio di noce.
Hawking Stephen	Lo sguardo rivolto alle stelle
Heckler Ricchard	Coincidenze
Heidegger, Martin	Che cos'è metafisica?
Heidegger, Martin	Il concetto di tempo
Heisenberg Werner	Fisica e filosofia
Hermann Greta	Fisica e filosofia
Hertog Thomas	Sull'origine del tempo
Hillman, James	Il codice dell'anima
Hoestatder, Dennett	L'io della mente
Hope Murry	IL segreto di Sirio
Hoyle Fred	L'origine dell'universo e l'origine della religione
Humphrey Nicholas	Polvere d'anima. La magia della coscienza
Hunter Tobias	Una nuova idea del mondo
Hurter Tobias	Una nuova idea del mondo
Jaffè Aniela	Sogni, profezie, apparizioni
James William	Esiste la "coscienza?
Jauch J. M.	Sulla realtà dei quanti
Jaynes Julian	La natura diacronica della coscienza
Jaynes Julian	Il crollo della mente bicamerale.
Johnson Steven	Dove nascono le grandi idee.
Joseph Frank	Il potere delle coincidenze.
Jou David	Riscrivere la Genesi.
Jung Carl	Psicologia dell'inconscio
Jung Carl	Esperienza e mistero
Jung Carl	Animus e anima
Jung Carl	Jung parla Interviste e incontri

Jung Carl	Il libro rosso. Liber novus.
Jung Carl	L'analisi dei sogni.
Jung Carl	La dimensione psichica
Jung Carl	L'uomo e i suoi simboli
Jung Carl	L'io e l'inconscio
Jung Carl	Tipi psicologici
Jung Carl	Ricordi sogni e riflessioni.
Kaku Michio	L'equazione divina
Kaku Michio	Il futuro dell'umanità
Kaku Michio	Iperspazio
Kane Gordon	Il giardino delle particelle elementari.
Kinnebriock Werner	Dove va il tempo che passa.
Kippenhahn Rudolf	Cosmologia da tasca.
Koch Christof	Sentirsi vivi
Koch Christof	Una coscienza. Confessioni di uno scienziato romantico
Koch Christof ok	La ricerca della coscienza
Kornelius Martin	Einstein Light
Kosko Bart	Il fuzzy-pensiero
Krauss Lawrence	Il cuore oscuro dell'universo. Alla ricerca della «Quinta essenza
Krauss Lawrence	L'Universo dal nulla
Kumar Manjit	Quantum. Da Einstein a Bohr, una nuova idea della realtà.
Kundera Milan	L'immortalità
Lalumera Elisabetta	Cosa sono i concetti
Lambert Karel, G. Gordon	Introduzione alla filosofia della scienza
Lanternari Vittorio	Religione magia e droga
Lanza R. + Herman	Biocentrismo
Lanza Robert	Il grande disegno biocentrico
Lanza Robert	Oltre il biocentrismo
Lederman -Hill	Fisica quantistica per poeti
LeDoux Joseph	Lunga storia di noi stessi
Ledoux Joseph	Il cervello emotivo
Licata Ignazio	Osservando la Sfinge.
Linden David J.	La bussola del piacere.
Lindley David	La luna di Einstein
Lipton Bruce	La biologia delle credenze
Livio Mario	La sezione aurea.
Lu Tung-pin	Il segreto del fiore d'oro
Magrini Marco	Cervello. Manuale dell'utente.
Marchesi Fabio	La fisica dell'anima
Marchesi Fabio	Exotropia

Marchi Vittorio	La scienza dell'Uno. Chiave dell'universo nascosto
Marconi Momolina	Preludio alla storia delle religioni
Marletto Chiara	La scienza dell'impossibile. Alla ricerca delle nuove leggi della fisica
Marraffa Massimo	Percezione, pensiero, coscienza
Mastrogiovanni G.	Un'introduzione elementare alla fisica quantistica
Mavaldi Marco	Le due teste del tiranno.
Mc Taggart Lynne	The Bond. Il legame quantico
Mc Taggart Lynne	La forza segreta dell'universo.
McAll Kenneth	Fino all radici
Mecacci Luciano	Identikit del cervello
Michell John	Astroarcheologia
Miramonti Reseghetti	Neutrino. La particella fantasma
Morowitz Harold	La nascita di ogni cosa
Mortier Roland	Impostori e creduloni nel secolo dei lumi
Motterlini Matteo	Trappole mentali.
Mugnos Sabrina	L'universo che pensa
Mullis Kary	Ballano nudi nel campo della mente
Musso Paolo	La scienza e l'idea di ragione....
Nagel Thomas	Una brevissima introduzione alla filosofia
Neumann Erich	Storia delle origini della coscienza
Nicolau Fausta	Costellazioni
Noe Alva	Perché non siamo il nostro cervello
Northoff Georg	Il codice del tempo. Cervello, mente e coscienza
O'Keane Veronica	Il bazar della memoria.
O'Regan J. Kevin	Perché i colori non suonano. Una nuova teoria della coscienza
Odifreddi Piergiorgio	Una via di fuga.
Oliverio Alberto	Dove ci porta la scienza
Ortoli - Pharabod	Metafisica quantistica
Pacori Marco	I segreti del linguaggio del corpo
Palazzi Pablo	La grande illusione
Palmarini e altri	Musica, luce, colori, la nuova energia
PanekRichard	Il secolo invisibile
Paolelli Ermanno	Neuro-quantistica. La nuova frontiera delle neuroscienze
Parigi Silvia	Magia e scienza nell'età moderna
Paris Giorgio	In un volo di storni. Le meraviglie dei sistemi complessi
Pauli Wolfgang	Psiche e natura

Pearson Carol S.	L'eroe dentro di noi
Peat David	Sincronicità. Un connubio tra materia e psiche.
Peat David	I sentieri del caso
Pecere Paolo	La natura della mente
Peiretti Federico	Il matematico si diverte.
Pellegrini Massimo	Fisica quantistica per principianti
Penrose Roger	La strada che porta alla realtà
Penrose Roger	L'Universo è ancora un segreto
Penrose Roger	La mente nuova dell'imperatore
Perconti Pietro	Coscienza. Lessico filosofico
Piazza Manuela – Pavani F.	Le neuroscienze cognitive
Pim van Lommel	Coscienza oltre la vita
Pizzuti Marco	Dall'uomo 2.0 alla matrice universale della coscienza.
Platone	Tutte le opere
Poli Eric Francesca	Anatomia della coscienza quantica
Polkinghorne John	Teoria dei quanti
Popper Karl	L'Io e il suo cervello
Poretti Giacomo	Un allegro sconcerto
Powell Arthur	Il corpo astrale e relativi fenomeni
Prattico Franco	Dal caos… alla coscienza
Privitera Agata	MBT La teoria di Myers-Briggs e le 16 personalità
Proietti Massimiliano	Relazioni cosmiche…Universo partecipatorio
Proverbio Edoardo	Archeoastronomia
Radin Dean	Menti interconnesse.
Randall Lisa	Passaggi curvi.
Ranucci Stefano	Fisica dell'altro mondo.
Reda Giuseppe	L'energia e le sue forme di esistenza
Reeves Hubert	L' evoluzione cosmica
Regge Tullio	Infinito
Renn Jurgen	L'evoluzione della conoscenza
Rey Olivier	Itinerari dello smarrimento.
Rezzolla Luciano	L' irresistibile attrazione della gravità.
Rhine Louisa	Psicocinesi. La mente domina la materia.
Richet Pascal	Una storia naturale del tempo
Romano Giuliano	Archeoastronomia italiana
Romano Giuliano	Orientamenti ad sidera
Rosati Giancarlo	Coscienza cosmica. Percorso verso l'illuminazione
Rossanese Emanuele	Che cos'è una particella

Rossi Daniela	Materia che sogna
Roth Reme	I cercatori di Dio
Rovelli Carlo	Buchi bianchi
Rowan Robinson	I nove numeri del cosmo.
Rucker Rudy	La quarta dimensione.
Russel Bertrand	Dizionario di logica fisica e morale
Russo Michele	La vita è un “caso” della materia?
Sacks Oliver	Il fiume della coscienza
Saggio Isabella	L’età, se esiste
Sale Musio Carla	Comunicare con chi non ha più un corpo
Salerno Vincenzo	Gli enigmi del cervello cosciente.
Salvati Marco	I neuroni di Dio
Santamaria Giando	ChatGPT revolution
Savoldi, Ceroni, Vanzago	La coscienza. Contributi per specialisti e non specialisti.
Schumacher Ernst	Una guida per i perplessi
Searle John	La mente
Searle John	Il mistero della coscienza
Serra Pietro	Dall’anima alla coscienza. Una incursione nelle pieghe della mente
Seth Anil	Come il cervello crea la nostra coscienza
Severino Emanuele	La filosofia antica
Shapira Haim	Lezioni sull’infinito
Sheldrake Rupert	La presenza del passato
Sheldrake Rupert	Le illusioni della scienza.
Sheldrake Rupert	La mente estesa.
Siegfried Tom	Il numero dei cieli. Una storia del multiverso e della ricerca per comprendere il cosmo
Sigman Mariano	La vita segreta della mente. Come funziona il nostro cervello quando pensa, sente, decide
Singer Charles	Breve storia del pensiero scientifico
Sissa ISAS laboratori	Caos e complessità
Solms Mark	La fonte nascosta. Un viaggio alle origini della coscienza
Sorensen Roy	La stanza delle meraviglie filosofiche
Sparzani e Panepucci	Jung e Pauli. Il carteggio originale:
Spilli (Manuali)	L’illuminismo
Stanzani Maserati	Coscienza e significato
Stapp Henry	Quantum Theory and Free Will. Springer International Publishing.

Stapp Henry	Mindful Universe
Stapp Henry	Mind, matter and quantum mechaniscs
Styer Daniel	Capire davvero la relatività. Alla scoperta della teoria di Einstein
Suenens Jozef	Il riposo nello Spirito
Sun-Tzu	L'arte della guerra
Surace Graziano	Il Quanto di sfida
Surprise Kirby	Sincronicità. Capire e utilizzare le coincidenze significative
Tagliagambe Malinc	Jung e il libro rosso: il Sé e il sacrificio dell'Io. Moretti & Vitali.
Tagliagambe Malinc.	Pauli e Jung. Un confronto su materia e psiche. Raffaello Cortina Editore.
Tagliagambe Silvano	Lo sguardo e l'ombra
Tagliagambe, Desideri, Vitiello, Pieri	Mente, cervello, ambiente: questioni
Teilhard de Chardin	L'avvenire dell'uomo
Teilhard de Chardin	Il fenomeno umano
Teillard De Chardin	Le singolarità della specie umana
Teillard De Chardin	Il posto dell'Uomo nella Natura
Teodorani Massimo	L'atomo e le particelle elementari. Macro Edizioni
Teodorani Massimo	Sincronicità. Il legame tra fisica e psiche. Da Pauli a Jung a Chopra. Macro Edizioni.
Teodorani Massimo	Tesla. Lampo di genio
Teodorani Massimo	Raccontare l'universo. Introduzione divulgativa all'astrofisica
Teodorani Massimo	La mente creatrice. Dall'universo fisico alla vita intelligente. Macro Edizioni.
Teodorani Massimo	Entanglement. L'intreccio nel mondo quantistico: dalle particelle alla coscienza.
Teodorani Massimo	Bohm. La fisica dell'infinito. Macro Edizioni
Teodorani Massimo	I grandi numeri celesti
Teodorani Massimo	Teletrasporto
Tesla Nicola	Le mie invenzioni. L'autobiografia di un genio
Tesla Nikola	Sull'incremento dell'energia umana
Tipler Frank	The physics of immortality. Doubleday.
Tonelli Guido	Materia. La magnifica illusione
Tonelli Guido	La nascita perfetta delle cose
Tonelli Guido	Tempo. Il sogno i uccidere Chronos
Toni Paolo	Scintille matematiche. Giochi e gare di creatività e logica

Tononi Edelman	Un universo di coscienza
Tononi Giuio	PHI Un viaggio dal cervello all'anima
Tutto (De Agostini)	Illuminismo
Vaas Rudiger	Hawking. Universo, spazio, tempo
Vassallo Nicla	Filosofie delle scienze
Vidali Paolo	Storia dell'idea di natura. Dal pensiero greco alla coscienza dell'antropocene
Vincenzi Lorenzo	Illuminismo.
Von Franz Marie-Louise	Psiche e materia
Von Franz Marie-Louise	Le fiabe interpretate
Von Franz Marie-Louise	Divinazione e sincronicità. Psicologia delle coincidenze significative
Von Franz Marie-Louise-ok	Tipologia psicologica. Le funzioni della coscienza: pensiero e sentimento, intuizione e sensazione
White John	L'incontro tra scienza e spirito.
Widmann Claudio	Sincronicità e coincidenze significative.
Widmann Claudio	Introduzione alla sincronicità. Edizioni Magi.
Wiener Anna	La valle oscura
Wigner Eugene	L'irragionevole efficaci della matematica nelle scienze naturali
Wilber Ken	Questioni quantistiche
Wilson Edward	L'armonia meravigliosa
Wolf Andrea	Il passaggio di Venere
Wolf Fred	Anelli temporali e torsioni spaziali.
Zeilinger Anton	Il velo di Einstein
Zeilinger Anton	La danza dei fotoni
Zichichi Antonino	L' irresistibile fascino del tempo.
Zichichi Antonino	L'infinito
Zolla Elemire	Archetipi. Aure. Verità segrete. Dioniso errante
Zoppi Giorgio	Cosmologia a sei livelli di difficoltà
Zukav Gary	La danza dei maestri Wu Li
Zukav Gary	Storie dell'anima. Storie vere che possono cambiare la vita

Series Manifesto.

We live in a unique era. Never before has science been able to describe the world with such precision, and never before has the world seemed so mysterious.

Quantum physics, which emerged in the early twentieth century as a response to technical problems in classical physics, soon proved to be something more: a ***revolution in the very way we conceive of reality****. It has not only transformed our knowledge but has also called into question categories we once believed to be untouchable: space, time, causality, objectivity, identity, and even the idea of "reality" as something independent of the observer.*

This series stems from a simple yet bold conviction: quantum physics does not concern only the microscopic world, but directly challenges the way we live, know, and make sense of existence.

Beyond popular science, without sacrificing rigor

These books are neither technical manuals nor academic treatises. At the same time, they are not texts offering spiritual escapism or easy consolation. The intent is different: to make the fundamental concepts of quantum physics accessible

- ***without mathematical formulas,***
- ***without oversimplification****,*

Each volume takes scientific knowledge—its history, its experiments, and its key figures—seriously. But it takes just as seriously the questions that science inevitably leaves unanswered.

When Science Meets Philosophy

Quantum physics has shown that the universe is not a predictable machine, but a ***field of possibilities****, where observation matters, uncertainty is structural, and the clear separation between subject and object breaks down. This gives rise to questions that are no longer merely scientific:*

- *What is reality really?*
- *Is there a world "out there" independent of the observer?*
- *What is the role of consciousness?*

- *Does chance govern everything, or is there a deeper form of order?*

This series explores these questions ***at the intersection*** *of physics, philosophy, and metaphysics—where no single discipline can proceed on its own.*

Metaphysics, Yes—But Without Illusions

The dialogue between quantum physics and metaphysics is fascinating, yet delicate. For this reason, each volume maintains a clear distinction between:

- *what is scientifically grounded,*
- *what is philosophical interpretation,*
- *what is symbolic or spiritual openness.*

You won't find miraculous promises or spiritual shortcuts disguised as science. Instead, you'll find hypotheses, analogies, and resonances, presented with caution and intellectual honesty.

From Theory to Everyday Life

Why should any of this matter to us, beyond mere intellectual curiosity? Because the worldview we adopt—often without realizing it—profoundly influences:

- *the way we deal with uncertainty,*
- *our relationship with choices and responsibility,*
- *our relationships with others and with ourselves.*

A deterministic universe gives rise to a certain conception of life. A quantum universe suggests another: one that is more open, more fragile, but perhaps also more creative.

Science and Inner Life

Without blurring the lines, this series goes so far as to explore the ***inner dimension*** *of the human being: consciousness, the experience of meaning, and spirituality understood not as dogma but as an openness to mystery.*

An Invitation to the Reader

The reader is not called upon to believe, but to reflect. Not to choose a truth, but to hold complexity and wonder together.

Back Cover

Can a cat locked in a box be both alive and dead at the same moment? The most famous image in modern physics continues to unsettle us because it is not merely about an imaginary experiment: it challenges the way we view reality, life, and ourselves. Erwin Schrödinger was one of the great architects of quantum mechanics, the theory that transformed our understanding of matter. But he was also something more: a physicist capable of questioning the meaning of existence, the origin of biological order, and the mystery of consciousness. Throughout his intellectual journey, his equations engage with biology, the observer problem with philosophy, and the scientific concept of unity with the ancient insights of the Upanishads and "Advaita Vedānta." This book tells the story of the thinker behind the paradox. It reconstructs the story of Schrödinger's cat, conceived not to celebrate quantum absurdity but to critique its most radical interpretations; it explores the question that gave rise to one of the most influential texts of the twentieth century, *What Is Life?*; and it traces the thread that links the order of living beings, genetic information, and the profound unity of nature. Without forcing improbable parallels between contemporary science and Eastern spirituality, the book compares two ways of questioning reality. On one hand, the rigor of physics, which measures, formulates hypotheses, and is open to correction. On the other, the wisdom of ancient India, which investigates inner experience and asks who the knower truly is. In between lies the problem of consciousness: are we mere spectators of the universe, or one of the forms through which the universe becomes self-aware? With clear language and accessible rigor, *Schrödinger and the Paradox of Life* guides the reader beyond simplifications about "spiritual quantum physics" and beyond the idea of a cold science, incapable of addressing the big questions. It is a journey through laboratories, revolutionary ideas, philosophical texts, and moments of contemplative silence. Because in Schrödinger's box, there is not just a cat. There is the unstable boundary between what we observe and what we are.

Synopsis ISBN

Schrödinger's cat is one of the most famous and misunderstood symbols of contemporary science. But behind the famous paradox lies an even deeper question: What is life, and what place does consciousness occupy in the universe? This essay portrays Erwin Schrödinger as a revolutionary physicist, a visionary biologist, and a thinker open to dialogue with Indian philosophy. From the enigmatic world of quantum mechanics to reflections on "What Is Life?", from the order of living organisms to the Upanishads and "Advaita Vedānta," the book explores the boundaries between matter, mind, and knowledge. With a clear style and accessible rigor, the author clarifies the true meaning of the cat paradox, avoiding simplistic notions of "quantum spirituality." A journey through science and philosophy that examines the role of the observer, the mystery of the self, and the idea of a profound unity of reality. A book for readers curious about physics, biology, philosophy, and the great questions of humanity.

The Author.

Bruno Del Medico is an Italian writer and science communicator, author of essays dedicated to the exploration of consciousness, metaphysics, and the new paradigms of quantum physics. He publishes with Edizioni PensareDiverso, a publishing house specializing in spirituality, critical thinking, and cutting-edge science. In his books, he explores the relationship between the observer and reality, proposing a vision of the universe in which consciousness is not a byproduct of matter, but a fundamental element of existence. Bruno Del Medico's approach integrates scientific concepts, philosophical reflection, and spiritual insights, presented in an accessible style aimed at fostering the reader's inner transformation.

He is considered a significant voice in the field of contemporary holistic outreach, particularly appreciated in alternative cultural contexts and on paths of personal exploration.

Edizioni PensareDiverso is an independent Italian publishing house specializing in non-fiction focused on new paradigms of thought, contemporary spirituality, and inner exploration, with a particular emphasis on topics at the intersection of science, consciousness, and metaphysics. The publishing house promotes works that challenge the traditional materialist worldview, explore the relationship between mind, reality, and the universe, offer holistic and systemic interpretations of existence, and foster personal and collective awareness.

Erwin Schrödinger - biography - quantum physics
Schrödinger's cat - quantum paradox
What Is Life? – Schrödinger – biology
quantum physics - consciousness - observer
Eastern philosophy - science - spirituality
Upanishads - "Advaita Vedanta" - non-dualism
science and philosophy - mind - nature of reality

Book on Schrödinger - popular science
Schrödinger's Cat Explained - Physics for Everyone
Quantum Mechanics for Italian Beginners
Consciousness and Physics - A Popular Science Essay
Schrödinger's "What Is Life?" with Commentary
The Upanishads and "Advaita Vedānta"—an introduction
Books on Science and Philosophy—Mind and the Universe

www.ingramcontent.com/pod-product-compliance
Lightning Source LLC
LaVergne TN
LVHW010713061126
851857LV00014B/1159